AF763137

Conservation Landscapes and Human Well-Being

The Himalayas are said to be the youngest mountain ranges in the world. This book studies the well-being of the Eastern Himalayan forest dwellers in terms of their capabilities and functioning. Using Amartya Sen's and Martha Nussbaum's capabilities approach, it examines the educational and health opportunities and substantial freedoms afforded to farmers and pastoralists living and working in the Senchal and Singalila Protected Areas of North Bengal, India. It also discusses the challenges and potential of the Forest Rights Act as a well-being delivery mechanism. The book adopts a comparative narrative of socio-ecological information generated from interviews, ecological field methods, remote sensing and participatory rural appraisals to provide insight on human development in conservation contexts.

This volume will be of interest to students and researchers of conservation biology, development studies, socio-ecological systems studies, political ecology, human development index, ecological economics, environmental sociology and South Asian studies. It will also be useful to policy makers and NGOs in the conservation and livelihoods sector.

Siddhartha Krishnan is Associate Professor at Ashoka Trust for Research in Ecology and the Environment (ATREE), Karnataka, India, and is affiliated to its Centre for Biodiversity and Conservation. He is the Convener of ATREE's Academy for Conservation Science and Sustainability Studies. His disciplinary and conceptual interests are in historicizing environmental sociology and sociologizing environmental history. He collects field and archival data to address questions pertaining to pastoral landscape and lifestyle transformations; human capabilities and ecosystem services; environmental justice; and modernity and development questions as they pertain to food, pesticide use and human health. He teaches sociology, environmental sociology and qualitative research methods in ATREE's PhD programme. He was Carson Fellow (2012–2013) at the Rachel Carson Center (RCC) for Environment and Society, Ludwig-Maximilians University, Munich, Germany. Between April 2015 and May 2019, he was a member of the Executive Committee of the Carson Society of Fellows. He was elected in April 2016 as a board member of the International Consortium of Environmental History Organizations.

Soubadra Devy is Associate Professor and former Co-Convener of the Centre for Biodiversity and Conservation at Ashoka Trust for Research in Ecology and the Environment (ATREE), Karnataka, India. Her research interests are in interactions between plants and animals, and developing a canopy programme for India through its Protected Areas network. She is also working on extending, through participatory community approaches, the biodiversity frontier to production landscapes. In 2006, for her treetop discoveries, she won the Lowell Thomas prize awarded by the Explorers Club and Rolex, USA. She also develops rigorous field conservation courses, which cater to various target audiences.

Neha Mohanty presently works as Education and Communication Officer at the Academy for Conservation and Sustainability Studies, ATREE, Karnataka, India. Her areas of interest comprise issues of violence and dispossession, local health traditions, sociology of agriculture and eco-social justice. She is also interested in exploring the impact on mental health of communities in stressful conservation contexts.

Transition in Northeastern India

Series Editor: Sumi Krishna
Independent scholar, Bengaluru, India

The uniquely diverse landscapes, societies and cultures of northeastern India, forged through complex bio-geographic and socio-political forces, are now facing rapid transition. This series focuses on the processes and practices that have shaped, and are shaping, the peoples' identities, outlook, institutions, and economy. Eschewing the homogenising term 'North East', which was imposed on the region in a particular political context half a century ago, the series title refers to the 'northeastern' region to more accurately reflect its heterogeneity. Seeking to explore how the 'mainstream' and the 'margins' impact each other, the series foregrounds both historical and contemporary research on the region including the Eastern Himalaya, the adjoining hills and valleys, the states of Arunachal Pradesh, Assam, Manipur, Meghalaya, Mizoram, Nagaland, Sikkim, and Tripura. It publishes original, reflective studies that draw upon different disciplines and approaches, and combine empirical and theoretical insights to make scholarship accessible for general readers and to help deepen the understanding of academics, policy-makers and practitioners.

Rethinking Economic Development in Northeast India
The Emerging Dynamics
Edited by Deepak K. Mishra and Vandana Upadhyay

Legal Pluralism and Indian Democracy
Tribal Conflict Resolution Systems in Northeast India
Edited by Melvil Pereira, Bitopi Dutta and Binita Kakati

Conservation Landscapes and Human Well-Being
Sustainable Development in the Eastern Himalayas
Edited by Siddhartha Krishnan
Associate Editors: Soubadra Devy and Neha Mohanty

For a full list of titles in this series, please visit www.routledge.com/Transition-in-Northeastern-India/book-series/TNI

Conservation Landscapes and Human Well-Being

Sustainable Development in the Eastern Himalayas

Edited by Siddhartha Krishnan

Associate Editors: Soubadra Devy and Neha Mohanty

Routledge
Taylor & Francis Group
LONDON AND NEW YORK

First published 2021
by Routledge
2 Park Square, Milton Park, Abingdon, Oxon OX14 4RN

and by Routledge
52 Vanderbilt Avenue, New York, NY 10017

Routledge is an imprint of the Taylor & Francis Group, an informa business

British Library Cataloguing-in-Publication Data
A catalogue record for this book is available from the British Library

Library of Congress Cataloging-in-Publication Data
A catalog record for this book has been requested

ISBN: 978-1-138-31397-2 (hbk)
ISBN: 978-0-367-53884-2 (pbk)
ISBN: 978-1-003-08359-7 (ebk)

Typeset in Sabon
by Apex CoVantage, LLC

Contents

Illustrations

Figures

Tables

Contributors

Annesha Chowdhury is a PhD student at the Academy of Conservation Sciences and Sustainability Studies, ATREE, Karnataka, India. Her research interests include how ecosystems and ecological processes underpin ecosystem services and human well-being of underserved minority communities. She takes special interest in how agro-ecosystems can be better managed to sustain biodiversity and people's well-being through a capabilities-oriented approach without impairing natural process and functions. Her current research focuses on the managed landscapes of Darjeeling tea plantations in the Eastern Himalayas, areas that lie outside of Protected Areas, and how they can contribute to support biodiversity while enhancing people's well-being in the landscape.

Sourya Das is a rural development and management professional and has worked mainly on ecosystem services, institutions and human well-being. Presently, he is working as Senior Researcher with Watershed Organisation Trust (WOTR) and is mainly associated with the work on economic valuation of natural resources. He has also worked with the West Bengal State Rural Development Agency (WBSRDA) for implementation of a food security program, and with Agricultural Finance Corporation Limited as a junior consultant in a water resource project. His research interests lie in the combination of economic and non-economic valuation of natural resources to establish robust linkages of ecosystem services and human well-being. He is also interested to work on the governance aspect of natural resources.

Soubadra Devy is Associate Professor and former Co-Convener of the Centre for Biodiversity and Conservation at Ashoka Trust for Research in Ecology and the Environment (ATREE), Karnataka, India. Her research interests are in interactions between plants and animals, and developing a canopy programme for India through its Protected Areas network. She is also working on extending, through participatory community approaches, the biodiversity frontier to production landscapes. In 2006, for her treetop discoveries, she won the Lowell Thomas prize awarded by the Explorers Club and Rolex, USA. She also develops rigorous field conservation courses, which cater to various target audiences.

Sarala Khaling is Regional Director, ATREE, Sikkim, and a wildlife biologist/ecologist and has worked in biodiversity conservation and ecosystem services at a landscape level, most of which was focused on the mountain areas of Nepal, northeast India and Bhutan. Her research interests include looking at drivers of change in human-modified landscapes that impact biodiversity, ecosystems and ecosystem services, especially in the mountains and hills of northeast India and North Bengal. She has also been working on building capacity of students, researchers, community leaders and local civil society organizations in northeast India.

Siddhartha Krishnan is Associate Professor at Ashoka Trust for Research in Ecology and the Environment (ATREE), Karnataka, India, and is affiliated to its Centre for Biodiversity and Conservation. He is the Convener of ATREE's Academy for Conservation Science and Sustainability Studies. His disciplinary and conceptual interests are in historicizing environmental sociology and sociologizing environmental history. He collects field and archival data to address questions pertaining to pastoral landscape and lifestyle transformations; human capabilities and ecosystem services; environmental justice; and modernity and development questions as they pertain to food, pesticide use and human health. He teaches sociology, environmental sociology and qualitative research methods in ATREE's PhD programme. He was Carson Fellow (2012–2013) at the Rachel Carson Center (RCC) for Environment and Society, Ludwig-Maximilians University, Munich, Germany. Between April 2015 and May 2019, he was a member of the Executive Committee of the Carson Society of Fellows. He was elected in April 2016 as a board member of the International Consortium of Environmental History Organizations.

Neha Mohanty presently works as Education and Communication Officer at the Academy for Conservation and Sustainability Studies, ATREE, Karnataka, India. Her areas of interest comprise issues of violence and dispossession, local health traditions, sociology of agriculture and eco-social justice. She is also interested in exploring the impact on mental health of communities in stressful conservation contexts.

Abhishek Samrat is Research Associate at the Eco-Informatics Lab at ATREE. His research interests lie in solving or highlighting ecological problems using Remote Sensing and GIS technology. He is presently working on peninsular Indian grasslands, Western Ghats rainforests and northeastern India forests.

Preface

A book's preface is its scope and aim. This organizing principle has a metaphysical appropriateness for a book on human well-being. The scope of this book then is the question of what scope remains for flourishing in the forest. Scope is traditionally about subject matter and opportunity. But in this volume, the subject *matters*. In conservation spaces, opportunity, along with choice, is resource along with natural resources. Both are crucial for human subjects to flourish. The aim of the volume then is not merely the purpose of the authors, but the purpose of human life in conservation landscapes. It is an adequately philosophized and empirically adequate study of well-being of Eastern Himalayan forest dwellers using Amartya Sen's and Martha Nussbaum's capabilities approach.

Scope: opportunities and choices to flourish in a forest

The book's prospects lie in how well it narrates what prospects remain for people farming the edges of the Senchal and Singalila Protected Areas (PA) of Darjeeling. A sense of these prospects is significant because the Forest Rights Act (FRA), as a well-being delivery mechanism, has not been implemented fully in the Darjeeling Himalayas or elsewhere in the Eastern Himalayas. Neither have *Panchayat* elections, which institutionalize local governance, been held for over a decade in Darjeeling. When implemented and held, the Act and the polls will form institutional contexts for individuals and families in forest villages to flourish. State, civil society and academic actors who would be involved in the implementation of rights, conduct of elections and post-entitlement and post-election monitoring will find this edited volume insightful in understanding what forest dwellers value. While policy processes rely on numbers, typically income and asset value, this volume and the capabilities approach it employs is a qualitative baseline of what people would like to be and do with rights and income opportunities. It is a quality-of-life primer in conservation contexts, with the Eastern Himalayas forming an immediate context. It engages with resources, in our case with rough equivalence, ecosystem services, that are crucial for life quality. These include provisioning and regulating services, or in resource terms, fodder, fuelwood, forest produce and water. Resources or ecosystem services are only instrumental to quality-of-life achievements and their mere presence does not signal quality in any way. What resources or services enable a forest dweller to be and do matters. The proof of the quality is in the living. The book scopes for what opportunities and choices remain in and around the two PAs, and what choices and opportunities farmers and their families value. Thus, we also invoke resources in the Rawlsian sense of public goods, and include rights, dignity, freedoms, roads, energy, schools and hospitals.

This edited volume is also an opportunity to push the boundaries of development or well-being scholarship in two senses – one on the larger sense of contextualizing or grounding the capabilities thinking of Amartya Sen and Martha Nussbaum in conservation spaces; an inter-disciplinary contextualizing and grounding, which is relevant to scholarship promoted in the 'Transition in Northeastern India' series of Routledge India. Published under the series is an edited volume, *Rethinking Economic Development in Northeast India* (Mishra and Upadhyay, 2017). In it a chapter by Nirod Chandra Roy deals with an idea that has been at the forefront of 're-thinking' development over the last two decades. The idea is 'Human Development'. Our four-year inter-disciplinary assessment of ecology and well-being in Darjeeling has afforded original insight on human development in conservation contexts. Two, we went beyond narrow well-being approaches typically deployed in socio-ecological assessments that concerned themselves with income as aggregate satisfaction indicator. We used the capabilities approach, more commonly referred to as the Human Development Approach. Human well-being for Nobel laureate Amartya Sen and the philosopher Martha Nussbaum is about capabilities or human functioning, that is 'being' and 'doing'; pursuing activities and achievements that one values. The key dignity and justice question for them is 'What is each person able to do and to be'? It is focused on choice or freedom, holding that the crucial constitutional services democracies should be promoting for citizens is a set of opportunities, or substantial freedoms.

Standard 'livelihoods' approaches account for, besides assets, people's capabilities in improving and sustaining their lives. Literature has it that the concept and definition of livelihoods is enhanced by the inclusion of quality-of-life philosophies like capabilities, which encompasses more than food or income and is about how people can function with their entitlements. In well-being monitoring that is locally sensitive and meaningful, capabilities is credited with potential to constitute a qualitative assessment. Despite livelihoods literature suggesting the suitability of the capabilities approach in assessing well-being, the approach has not been conceptually discussed, let alone been empirically operationalized, in conservation contexts, specifically forests. Our inter-disciplinary volume seeks to reconcile the multi-dimensionality of forest ecosystem functioning with the multi-dimensionality of human functioning in the Eastern Himalayas. The comparative assessments, conducted in seven villages near Singalila National

Park and Senchal Wildlife Sanctuary, involve rigorous biodiversity and water assessments, as well as capabilities-based well-being assessments.

Aim – the purpose of a forest life

The aim of the volume is more than just the purpose of this volume's team. It is about purposing human life in conservation landscapes. Less philosophically and more immediately, the aim of this book is to approach the well-being of Eastern Himalayan forest dwellers in terms of their capabilities and functioning. There are some larger aims, such as to assess and impart agency to forest dwellers and not just aggregate and infantilize forest communities as needy recipients, but as active agents. Also, the aim is to start a dialogue between 'community'-oriented conservation scholarship and the philosophically oriented capabilities scholarship. Here emerges a scope to mutually address critiques of both community and capabilities scholarship. The former is accused of valorizing community and ignoring individual choice. The latter is accused of ignoring community and valorizing the individual. Speaking of critiques, this volume also aims to contextualize not just the merits of the capabilities approach but also weaknesses. It locates the varied critiques of the capabilities approach as an 'over'- or 'under'-specified theory, or a universalizing or individualizing theory, in conservation contexts.

The volume aims to initiate another related dialogue between political ecology and capabilities philosophy – a dialogue between political and philosophical genres that find common ground in social justice and dignity. The capabilities scholarship values material and political entitlement that when contextualized in conservation settings maps on to political ecological concerns over land, conservation and decision-making rights. Political ecology can benefit from a philosophy of well-being courtesy of the capabilities school. Capabilities scholarship can well do with the politicizing of, for instance, attempts to confound and conflate animal rights and wildlife conservation.

Having said that, what this book does not aim to do is be a comprehensive review of forest livelihoods or well-being literature. The natural and social science scholarship around conservation and livelihoods is way too vast for a philosophy grounding volume to exhaust. Neither does the book pretend to be the alternative well-being perspective on conservation and livelihoods. It is, as mentioned already, a conversation starter on well-being in conservation contexts. Also, given the grounding of the capabilities philosophy in forests, the book does not attempt to exhaust the voluminous literature on capabilities in development contexts. The book does not exhaust the Singalila and Senchal regions either, let alone being truly representative of well-being challenges and opportunities in the whole of Darjeeling. The seven villages in which we conducted our ecological and well-being research were, however, carefully chosen to be representative of ecological, economic, political and cultural realities and diversities that characterize Darjeeling rurality.

In similar vein, it is impossible to *exhaust* all whose time and effort contributed to this volume. The team acknowledges a range of individuals and institutions with apologies to anyone who may feel left out. In no particular order, the team thanks all the residents of the seven villages in and around Singalila National Park and Senchal Wildlife Sanctuary. They spent valuable time with us discussing things they value. In the laborious and busy rural-scape that is the Himalayas, being interviewed for hours is an act of much patience, grace and decency. The porters who carried kilos of middle-class luggage, not to mention scientific equipment, during fieldwork deserve applause and appreciation. However, this is not an attempt to valorize their lives. They are tiring and thankless.

Thanks to the field assistants of ATREE, Darjeeling and Sikkim. ATREE's Eco-Informatics Lab produced some valuable remote-sensed imagery for the study. We thank the lab for its services. The team is grateful to the Accounts Department of ATREE, Bengaluru, Darjeeling and Gangtok for its financial servicing of the research project on which this book is based. The project 'Integrated Approaches for Adaptive Resilience Based Management of Forest Agro-Systems in the Sikkim-Darjeeling Himalayas' was funded by the Tata Social Work Trust (TSWT), a wing of the Tata Trusts. The Tatas have been patient and generous donors of ATREE. This is our sincere acknowledgement of their support. We thank the Wildlife and Forest Departments of West Bengal and Sikkim for the research permits. Their cooperation is only reflective of a more progressive disposition, in comparison to the rest of India besides Kerala, in caring and planning for forest dweller well-being. The team is grateful to Breena Holland of Lehigh University. She exposed us to the community of capabilities scholars the world over. Her support in getting the team to Cape Town in 2017 for the Capabilities Conference, where we shared some of the ideas that appear in the volume, was crucial. Finally, the team would extend its thanks to series editor Sumi Krishna, whose faith and patience in our work and this edited volume are what made it possible. The volume owes much to her. Shoma Choudhury, Commissioning Manager at Routledge, has been as patient and encouraging of our effort.

1 Human well-being in conservation landscapes

Siddhartha Krishnan

The human subject in conservation landscapes

Europe deforested and afforested its colonies. An institutionalized process, tree felling and plantation afforestation occurred under a famed but fraught structure of scientific forestry. Decolonization increased deforestation and afforestation in ex-colonies. In independent India tree felling and forest logging continued as a commercial and industrial enterprise. 'Deforestation' did not so much convey ecological impact, as it did of environmental impact. The 'crisis' discipline of conservation biology that emerged in the early 1980s conveyed the ecological consequences since the mid-1980s and early 1990s through the idea of biological diversity (biodiversity) and its loss. Biodiversity included genetic, species and ecosystem diversity and this diversity's significance for human well-being has never been understated since the 1992 Convention on Biological Diversity (CBD). Well-being lay implicit in the CBD's mission of conserving biodiversity, the sustainable use of its components and fair sharing of genetic use benefits. A philosophy of ethics is involved in conserving biodiversity. The act of conserving is an ethical one. Conservation, sustainability and equity are all 'consequentialist' – such actions are assumed to result in or have well-being consequences. But well-being can also involve a 'de-ontological' ethic, that is these actions are right by themselves and need not involve any calculation of consequences. Conservation has remained more of a consequentialist field inasmuch as human well-being is concerned. While deep-ecological ideologies that insist on preserving faunal species for their intrinsic worth has de-ontological implications, the conservation field has remained consequential for humans in material and philosophical ways.

The conservation landscapes of the volume are Protected Areas (PA). These include National Parks and Wildlife Sanctuaries in the Darjeeling Himalayas. We assessed well-being and ecosystem services in two Protected Areas, Singalila National Park and Senchal Wildlife Sanctuary. Humans dwelling in or near forests and dependent on them have been state, research and human rights subjects. As state subjects, forest dwellers have been less of a development constituency than a conservation constituency. The latter involves subjection to conservation laws that regulate or restrict resource access. Forest and wildlife laws thwart generation of agricultural, grazing or minor forest income. Income-spending opportunities and choices are usually curtailed by physical inaccessibility to markets. Earning and spending can be interpreted as freedoms under the right to life enshrined in the Constitution. Forest dwellers may well value allopathic treatment in hospitals and want to spend part of any income they manage to earn. However, absence of roads curtails such crucial expenditure. Roads are a subject matter of the Constitution's 11th Schedule. Many development components of this schedule and the delivery institution of the *Panchayat* stop at the fortress or forest gates of PAs. Their fates as subjects of the conservation state since independence has made them subjects of human rights activism and critical conservation scholarship since the late 1980s and early 1990s. Dignity and justice were the foundations of off-line and online protests and lobbying for entitlement in the early years of the new century. The legislation of the Recognition of Forests Rights Act (FRA) in 2006 was an entitling watershed. The pre– and post–forests rights discourse has been just but limiting for a well-being philosophy. This is because, residing in forests along with people and wildlife, is Jeremy Bentham's ghost. His utilitarianism haunts conservation policy and practice. The haunting is in two acts, one as public good, and two as good taken for granted.

PAs have been seen as a greater common albeit biologically diverse good. 'The greatest happiness of the greatest number' is a 'good' maximizing philosophy of Protected Area policy. This includes any bio-centric or deep-ecology strain of animal happiness. In this aggregation or sum-totalling of happiness, the distribution of pleasure is ignored. Forest dependents and dwellers suffer restriction, regulation and possible dispossession. They are a happiness collateral – they must be happy for others; others' happiness is theirs. Injustice here is not just a matter of collateral damage. Injustice or an unjust society for utilitarianism is a state of affairs where there is a loss of aggregate utility or happiness. However, injustice lies not just in an unfortunate state of affairs where a minority suffer for the majority. It also lies in people taking pleasure in others' pain. Powerful conservation elites promoting wildlife protection for their pleasure and also ostensibly the pleasure of animals actively seek to thwart the well-being prospects of forest dwellers. Structurally, by opposing the Forest Rights Act, and more specifically by lobbying for, say, the restricting of roads or fuelwood collection. Or besides this 'in-situ displacement', the active lobbying for eviction and displacement.

Utilitarianism also unfortunately haunts conservation activism and scholarship critical of forest citizens becoming collateral to the greater public good principle. Theirs is a well-intended but inadequate belief that entitlement with individual and community rights leads to satisfaction or pleasure. Rights mitigate the status of forest dwellers as a precariat. It can also attenuate income generation deprivations. Essential as these public goods are, that is dignity, income and liberty, they do not epitomize or exhaust well-being. That is as an end in themselves. This is a welfarist position that evaluates a state of affairs, in this case the entitlement, on the basis of utilities of that particular state. As much as these public goods, especially dignity and liberty, are also a well-being end by themselves, they also need to be evaluated in terms of choices and opportunities that a forest dweller can afford to pursue ways of living, including non-forest. Alternatively, they can be evaluated within the 'traditional' community confines of a forest village and pursue modern ways of thinking and living. Utilitarianism may not attribute any intrinsic significance to rights and liberties. These gain value in the calculus only to the extent they influence utilities or usefulness. Community activists and academics may differ on this count and actually attribute intrinsic value to rights and freedoms. However, freedoms have inherent worth by themselves (being free is a well-being constituent) and for the choices and opportunities they offer people to be and do things they value. The freedom to develop is important. After that, development as freedom is an idea that must be entertained democratically. This book seeks to release Himalayan forest dwellers and dependents from being subjects of an inadequate discourse of well-being. It makes a case for how the capabilities approach, or substantive freedoms and opportunities to be and do different valuable things, is a conversation starter in the conservation arena.

This edited volume is a critique of conservation strictly tethered to utilitarianism. By utilitarianism we mean 'the claim that all choices (of actions, rules, institutions, and so on) must be judged by their consequences, that is, by the results they generate'. We need distinguish 'between "culmination outcomes" (that is, only final outcomes without taking any note of the process of getting there, including the exercise of freedom) and "comprehensive outcomes" (taking note of the processes through which the culmination outcomes come about)' (Sen, 1999, p. 29). Attempting and achieving sustainable conservation by fiat or heavy regulation is one thing, and doing so with free choice and opportunity is quite another important thing. Well-being is not about generating and spending forest incomes in restricted or regulated ways. It is about the forest dweller having and exercising opportunities and choices democratically and valuably. Further, virtues like cooperation or sustainability, and services like regulating soil erosion or provisioning leaf litter, are not instantly and constantly 'means' to well-being. They characterize and constitute well-being itself. In these ways, the volume reconfigures and reinvigorates well-being in conservation landscapes.

The volume also rescues well-being from a taken-for-granted end state that it suffers in conservation discourses, however well-intended they are. The material or well-being consequences of 'fortress' conservation are well known (Brockington, 2002). Forest and pastoral communities suffer resource and income deprivations. But there are philosophical implications of conservation discourses, including of progressive human rights variants. A certain well-intended presumptuousness prevails. Namely, entitling communities or discussing their 'serviceability' by ecosystems processes and function has inevitable well-being benefits. Here, well-being is defined away as a mere utilitarian matter of income generation and satisfaction or ecosystem serviceability and satisfaction. Rights and services will generate income, and income will generate pleasure is the consequential sequence professed in conservation and livelihoods discourses. Natural resources, ecosystem services and income are of material and instrumental significance to human well-being. But well-being suffers relegation as a general 'satisfactory' end. Conservation discourses appear to close off any philosophical conversation on well-being that has practical and policy purchase. Not letting an indigene perish in a forest is just and good. But enabling the indigene to flourish in valuable ways is as just and fair.

Human well-being or flourishing was a moral question that engaged Aristotelian philosophers. Purposeful pursuit of happiness by cultivating virtues was an individual's craft. However, happiness became a matter of statecraft. Since World War II, welfare stood in for well-being. The care-needing citizen replaced the virtuous agent. The decolonized nation as body sovereign became grand political narrative. The human agent's body *as sovereign* remained a philosophical footnote. Along with citizen's legal rights, legal obligations dominated moral obligations. Citizen welfare became the state's political and legal obligation. The 'individual's' moral 'obligation' to pursue true potential was not a parliamentary agenda. The state was wary of leaving well-being to individuals or communities.

There was a Tocquevillean 'snag' as Bauman wrote (Bauman, 2000, p. 36). De Tocqueville

> long suspected, setting people free may make them indifferent. The individual is the citizen's worst enemy. The 'citizen' is a person inclined to seek her or his own welfare, through the well-being of the city – while the individual tends to be lukewarm, sceptical or wary about 'common cause', 'common good', 'good society' or 'just society'. What is the sense of 'common interests' except letting each individual satisfy her or his own? Whatever individuals may do when they come together, and whatever other benefits their shared labours may bring, portends constraint on their freedom to pursue what they see fit for each separately, and will not help such pursuit anyway.
>
> (Bauman, 2000, p. 36)

But a common cause indeed constrained the freedoms of individuals, the common cause being poverty. The wretched lacked Rawlsian goods like rights and income, goods that are means for flourishing or partial expressions of it (Chambers, 1995). Colonization and World War II saw the concentration of wealth and revenue; concentration in the colonizer's home coffers, or within the colony, in military coffers. Decolonization coincided with the war ending, and citizen welfare gained politico-moral ascent. Before we discuss state welfare, a philosophical detour on individual well-being is foundational. This is also an opportune moment to signpost this chapter's scheme for the reader.

Chapter scheme

The capabilities approach has Aristotelian roots. We roundup neo-Aristotelian work on human flourishing, obligation, virtue and potential. This we follow with a historical discussion on war, post-war and post-colonization development. We round up relevant political sociological and historical sociological work on development and welfare. Post-decolonization development policies had unintended socio-ecological consequences. Poverty, pollution and degradation intensified as did resource conflicts. This happened even as class and caste elites hoarded wealth. We discuss ecological consequences through environmental historical literature.

Poverty reduction and eradication remained a core development concern. Income and wealth inequality persisted into and intensified in the globalization years. Poverty and the narrow obsession with economic growth deepened and so did critiques of how the state need measure development and poverty. The capabilities approach was one such critique. Capabilities proceeded parallel to, in theory if not in rhetoric, the development-poverty debates of the 1980s and 1990s. A qualitative measure of freedom per person, it was a corrective to quantitative per capita measures. Capabilities gained institutional heft through the Human Development Reports since 1990. But it was also gaining *indicative* heft in another critical field which was livelihoods. We review 1990s livelihoods literature for its well-being philosophy and how the capabilities theory serves as an indicator of sustainable livelihoods. The review includes biodiversity conservation which became central to sustainable livelihoods since 2000. A critique of the well-being philosophy of the Ecosystem Services (ESS) framework follows. Promoted by the Millennium Ecosystems Assessments (MA), it suggests an elaborate services-human relationship.

We then make a strong and novel case for adopting capabilities in assessing and informing well-being in conservation landscapes. This includes a brief overview of Amartya Sen's capabilities approach, Martha's Nussbaum's capability list and Breena Holland's environmental extension of Nussbaum's list. Finally, we describe the Eastern Himalayan landscape, its ecology, economy and society. The chapter ends with summaries of chapters in the volume.

Human well-being: flourishing, obligation, virtue and potential

Aristotle's philosophy on human flourishing or self-realization has influenced the capabilities approach (Brown and Westaway, 2011). Sen leaves vague his conception of a good life. But Nussbaum endorses an Aristotelian conception of human flourishing, one that is also Marxian as she prefers, and not Marxist. Karl Marx's philosophical conception of human nature is near Aristotelian. For Nussbaum, the founding intuition behind the capabilities approach is moral.

Humans exert a moral claim that politics develop their capabilities. With right educational and material support, humans become capable of optimal functioning. This intuition draws from Aristotle's view that human functionings and capabilities need to influence political arrangements. The intuition is also Marxian. It is immoral that circumstances reduce humans to function in inhuman ways.

So what is it to be human? What is it to function and flourish as a human? For rhetorical and clarificatory contrast, let us pitch Aristotle against Jeremy Bentham, that is an essential versus utilitarian philosophy. Further, as we will see, a 'balancing' versus 'avoidance' prescription. In demarcating happiness as individual and social states, both clarify the political state's role. Human happiness suffers a conventional 'welfarism' interpretation in development and sustainable development discourses, thanks to Jeremy Bentham and, his student, John Stuart Mill. Let us discuss utility in brief and proceed to purpose.

English jurist and reformist Jeremy Bentham inspired traditional welfare economics. His famous and fundamental axiom was 'the greatest happiness of the greatest number'. The government's purpose was ensuring this principle. Utilitarianism is an ethical theory because it posits a moral principle – the happiness dictum. It is a *consequentialist* theory or one that judges an action's moral consequences. For utilitarianism, common happiness is the required moral consequence. But there was a problem with how happiness is distributed. For Sen, Bentham's concern was 'the total utility of a community', and not the 'the distribution of that total' (Sen, 1999a, p. 59).

There was in this 'an informational limitation of considerable ethical and political importance'. In the 'utilitarian ideal world', a handicapped person suffers 'a lower share of the given total'. She has 'lower capability to generate enjoyment and utility out of income'. For Sen the 'unifocal priority' of 'maximising the sum-total of utilities' has 'peculiar consequences' that justice philosophers like Rawls critiqued (Sen, 1999b, p. 352). For Rawls, every 'person possesses an inviolability founded on justice that even the welfare of society as a whole cannot override' (Rawls, 1971, p. 3). Greater good cannot right 'loss of freedom for some'. The 'rights secured by justice are not subject to political bargaining or to the calculus of social interests' (Rawls, 1971, p. 4). Welfare economics adhered to the utilitarian calculus through the 19th century. In the early 20th century it felt the heat. Sen (1999a) says masterful justice critiques like that of Rawls could have ensued as utility summing trumped distribution. Yet, science trumped justice as critique. Economists influenced by logical positivism critiqued utility as unscientific. Interpersonal comparisons for utility were difficult. Sen quotes economist Lionel Robbins – 'Every mind is inscrutable to every other mind and no common denominator of feelings is possible' (Sen, 1999a, p. 67). In embracing utility, late 19th- and early 20th-century economics embraced psychology. Pleasure was the operative and hedonistic noun. Hands (2009) provides an interesting history of the relationship between welfare economics and psychology, the competition between cardinal and ordinal measures of satisfaction and the prevailing of utilitarianism in welfare policy through preference and choice.

So if utilitarianism prevails in welfare policy, what of the justice critique? The one that Sen rues did not happen? Well if it had, the fairness of the calculus would be in question, and not its rigour. For, a Rawlsian critique of utility is about the quality of life, and not the quality of data. Rawls's is a 'philosophical conception for a constitutional democracy'. For him 'justice as fairness' is a 'systematic alternative to utilitarianism' (Rawls, 1971, p. xi). His primary goods did not depend on psychology for their status. For a thing to be a primary good, it depends on a person's morals as a citizen. Primary goods are what 'persons need in their status as free and equal citizens, and as normal and fully cooperating members of society'. These goods meet people's needs as citizens as opposed to their preferences and desires (Rawls, 1971, p. xiii).

For Rawls, his theory tallies with the Aristotelian tradition. Refrain from 'pleonxia' is the sense Aristotle assigns to justice and the familiar formulations deriving from it. Pleonxia is gaining advantage by seizing another's property, reward and office; denying one's due; disrespect and the like. This definition has an action framing. A person is just if by character she has an 'effective desire to act justly' (Rawls, 1971, p. 10). Aristotle's definition is an account of individual entitlement, that is what belongs and is due to a person. Rawls believes entitlements derive from institutions and legitimate expectations they generate. His definition of social justice applies to the basic structure or institutional protocols. These include the distribution of fundamental rights and duties, and the determination of advantages arising from social cooperation. So, Rawls's institutional critique of utilitarianism ties with an individualist tradition of Aristotle. His is a liberalism that is Aristotelian for its insistence on choice. In fact, for Nussbaum, Rawls's justice theory is a 'major liberal view' in the Aristotelian tradition of choice. Nussbaum maintains that Rawls's liberal view is Aristotelian for his insistence that choiceless satisfaction is of no moral worth (Nussbaum, 1992, p. 225). What then is the Aristotelian tradition of human well-being?

Human flourishing: the Aristotelian tradition

Rawls finds in Aristotle a perfectionist tendency. Perfectionism is at play in the realization of human excellence in various cultures. If we define good as pleasure, 'we have hedonism; if as happiness, eudaimonism' (Rawls, 1971, p. 22). An essentialist and subjective perspective of well-being is implied here. Rasmussen asks if there is a link between our understanding of human flourishing and human nature. Can one read-off flourishing's content from human nature? Is there no place for diversity and individuality (Rasmussen, 1999, p. 1)? Let us discuss essentialism's critique and the case for Aristotelian essentialism. Then we discuss, with hypothetical Himalayan examples, the features of Aristotle's objective and agentic conception of flourishing.

Essentialism and well-being

Nussbaum writes about essentialism being a 'dirty word' in academia. Taking essentialism to mean 'human life has central defining features', she examines critiques (Nussbaum, 1992, p. 205). These include an insensitivity to women and minority voices in addition to being in league with sexism, racism and patriarchy. Extreme relativism is a 'recipe for social progress' (Nussbaum, 1992, p. 205). But Nussbaum grants it to some criticism forms. They redirect attention from 'unexamined abstract assumptions to the world and its actual history' (Nussbaum, 1992, p. 205). Without some kind of essentialism, compassion and respect will evade the world. These two moral sentiments are necessary for a decent life. Nussbaum's Aristotelian essentialism is 'internalist'. A way of looking at and thinking about ourselves. She eschews any metaphysical foundations for what is indispensable for human life. What without which a human life does not count as one, she asks. Basic functional capabilities, then, is a 'higher threshold'. Falling beneath it does not make for a 'good human life' (Nussbaum, 1992, p. 221). We discuss these capabilities in a dedicated section later in the chapter. Suffice to say these capabilities are the goals of planning and legislation. For an Aristotelian essentialist, a life deficient in any one capability lacks humanness. She also needs to meet the objections of essentialism. These are neglect of historical and cultural difference and neglect of autonomy. Capabilities are general and fundamental; they allow for multiple specifications. However determinate be the conception of good, a latitude for citizens to individually and traditionally specify remains. Local specification entails sensitivity to context, human character and social relations. As for agency, the Aristotelian's capabilities list is not about actual functioning. There is emphasis on choice. The state does not nudge citizens to act in 'valued ways'. Instead it fosters conditions and circumstances for citizens to act in these ways. Choice matters. Choice, as we shall see, is also written into the essential capabilities list. The capability of choosing is fundamental to human essence. The Rawlsian-liberal view also converges with Aristotelian essentialism. Rawls holds that satisfactions that do not result from one's choices have no moral value (Nussbaum, 1992, p. 225).

Aristotelian flourishing

Since the 1980s, human flourishing gained ethical significance. The traditional translation of 'eudaimonia' as happiness did not convey its 'objective' meaning. For neo-Aristotelian ethicists like Douglas Rasmussen, flourishing is an alternate to consequentialist and de-ontological theories. Consequentialistic theories judge actions by their positive or negative consequences. Sustainable fruit harvest, for instance, is good because it is good for ecology and future humans. De-ontological theories judge an action's morality by the rightness of the act. Duty and right are basic, and an act's consequence is not considered. Sustainable honey harvest is a duty. But morality needs explaining and justifying. Because morality is not 'self-justifying', there is a difference between theoretical and practical knowledge. Morality has a function for flourishing. Flourishing always involves the contingent and particular. Practical wisdom, Nussbaum's architectonic capability, is for Rasmussen a 'central intellectual virtue' (Rasmussen, 1999, p. 19). Practical knowledge is not means-end rationality. It is the actor's ability in contingent circumstances to discern what is moral. One can only weigh well-being goods and virtues using practical knowledge. It takes practical wisdom to achieve these choice-worthy goods and virtues especially considering generic, individual and circumstantial factors. The six features of flourishing that Rasmussen lists follow and are illustrated through Eastern Himalayan instances.

1. Flourishing is objective. It is choice-worthy and desirable. Its very constitution makes it good. So, it is ontological and a state of being; it is not a feeling. An actuality or end, flourishing is action and activity. Activities both express and produce actualization of human potential. Participatory forest conservation or sustainable fuelwood collection do not lead to well-being. They are well-being themselves.
2. Flourishing is an inclusive end. It is not an end-all and goods and virtues reduced to mere means. Rawlsian primary or basic goods like health and education, and virtues like courage and justice are valuable not as means to flourishing. They are 'partial realizations or expressions' of flourishing (Rasmussen, 1999, p. 4). Flourishing is also an inclusive end in another sense. One can pursue activities as both means and as valuable things themselves. A *Lepcha*'s justice mindedness and education are expressive traits of her well-being. She can sow a diverse kitchen garden for nutrition and for present happiness.
3. Human flourishing is individual and diverse. Flourishing, goods and virtues may sound abstract and universal. But flourishing expressions and meanings differ between people. A *Lepcha* and *Rai* actualize their potentialities differently. Potential is general and individual. Individuals are not passive repositories for generic goods and virtues. When an 'individual's particular talents, potentialities, and circumstances' jointly engage, 'goods and virtues become real or achieve determinacy' (Rasmussen, 1999, p. 6). An ecosystem service and its sustainable use become determinant depending on the *Chettri*'s geographical, caste and class locations.
4. Flourishing is agent relative. People are not mere placeholders for flourishing. Flourishing gains value only in relationship with a person's life.

A person's activities constitute flourishing. The *Sherpa* is no passive recipient of ecosystem service flows. A service gains value in relation to *Sherpa* activity, for instance, the labour he expends in accessing water and tending to soil. The contrary aids in realizing the import of this relation. Consider agent-neutral impersonal ethics. It holds that values, reasoning and rules are abstract and universal. John Stuart Mill's utilitarianism is one such ethic. Here, forest edge farmers are placeholders. The *Rai* can substitute the *Chettri*. We celebrate Himalayan indigeneity, but ignore caste, class and power inequities. The particular and contingent are of ethical value too as are social and cultural differences. A person has real potentialities and circumstances. These 'characterize both what and who a person is' (Rasmussen, 1999, p. 9). But how can a good or virtue be both self-valuable and agent-relative? The value of goods and virtues is not a matter of their being mere means to human flourishing; rather, it is a matter of their being expressions or realizations of it. The diverse, terraced farms amidst Gorkhey's forests express peasant well-being. Yes, the barley crop is a nourishing means. But its cropping is a realization of valuable peasant activity.

5 Flourishing is self-directed activity. Possession and use of good is not flourishing. A person need take charge of her life and cultivate virtues that help in attaining goods. Senchal's villages may have a moral economy of reciprocity. In it must *Chettris* take part with honour and fairness. Seeking knowledge, cultivating character, choosing well and right action make for 'moral excellence'. 'Virtuous living is anything but passive' (Rasmussen, 1999, p. 11). Ethical forest dwelling involves a purposive and active morality. Practical reasoning is an activity that requires such effort. Reasoning is not automatic. The *Lepcha* must exert to forage leaf litter and find value in litter sharing with aged families.

6 Humans are naturally social. We mature with others. Some of our potentialities are 'other-directed' that we must actualize. Goods and virtue possession is not exclusive and selfish. We originate and mature in some social context, say community. Our values and sense of self emerge in such contexts. Human lives intertwine and we flourish 'with and among others' (Rasmussen, 1999, p. 13). But we need not be selective in our sociality. We can engage with those with whom we do not share common values. The *Lepcha* owes it to his Buddhist community for his social nature. He flourishes amongst his folk. But he is not obliged to not open a virtuous relationship with Hindu *Chettris*.

We have heard of Indian kingdoms flourish. But did individual citizens and subjects flourish? Ancient Indian or Moghul historians must tell us. We are exploring capabilities of today's forest dwellers here. But like the rest of their rural counterparts, they suffered under the crown and company. *Adivasi* and *Savarna* forest dwellers suffer what Sen terms as capability deprivations. We will get to such 'conservation' deprivations soon. But first the political sociology and economic history of war, decolonization and welfare. The three are a political-economic assemblage. We will disassemble and reassemble these in brief. The following is also an environmental history of welfare in general, and of well-being deprivations in conservation landscapes in particular.

World War II, decolonization, poverty and welfare

India was never at war. Neither of the world wars had us fighting out of our own will. We soldiered, died and survived for the British. The world wars, especially the second one, played important political and economic roles. Decolonization and welfare statism ensued. Plentiful environmental histories critique and chronicle the colonial fates of India's natural resources. The British commercialized forest spaces and species for railways and shipping. *Adivasis* experienced capability deprivations as forests ceased to be commons. Colonial commerce was capitalist, extractive, monopolistic and profitable. But there were war runs in this unjust and unsustainable market. Military markets emerged during wars. World War II's end ushered independence in India and some other colonies. But what were the specific roles of the wars in the empire? What did these wars do to resources and welfare in India?

India's colonial history was a capitalist one. The company, crown and cronies profited. Criticism of Marx and Marxist criticism has dominated discussions on colonial capitalism. Marx went beyond the nation and configured capitalist evil as an international system. He also configured good in capitalism, that it was a necessary evil. Critiquing British colonialism, he 'described its long-term impact as positive'. It 'would unify the Indian state and revolutionize the country's moribund economic, social, and political structures' (Steinmetz, 2014, p. 86). India unified, but the peasantry and working classes did not exactly flourish. Industrialists and landlords benefitted from unsustainable state policies. Liberals and leftists rejected the 'necessary evil' thesis. Imperialism or monopolistic finance capitalism, they argued, devastated colonies. Marxists also expounded under-development, dependency and world systems theories that critiqued capitalism. But primitive accumulation is of immediate relevance. Accumulation here is any capital or material stock upon which capitalist exploitation of labour can happen. Political ecologists critiquing power relations in a PA conservation invoke 'accumulation by dispossession'. For David Harvey, 'accumulation by dispossession' is not a simple historical starting point for capitalism. It is a recurring process that often takes an imperialist form. PA laws dispossess forest dwellers in- and ex-situ. Here, one can talk of capability displacements like of deprivations (Harvey, 2005, p. 159).

Generally, wars displace and dispossess. World War II is to have dispossessed the empire of its colonies. Warfare and post-war economic downturn in European economies hastened decolonization not to discount passionate independence struggles such as India's. But post-independence state welfare did not improve quality of Indian lives, especially that of rural India. Gadgil and Guha historicize how and why economic poverty and environmental degradation ensued (Gadgil and Guha, 1992). Before we discuss post-independence welfare failure and environmental decline, a discussion of warfare and welfare is instructive.

Historical sociologists provision an interesting insight on welfare's warfare origins. For Andreas Wimmer, many studies locate the welfare state's origins in war-veteran caring. The modern welfare state's institutional foundations lay in veteran care. For Wimmer, Theda Skocpol's (1992) work on social policy's political origins is a foundational instance. Her historical sociology identifies caring for veterans and their families as an 'institutional innovation'. For Europe, Wimmer cites Klausen's study of how nations transformed World War II economies into welfare economies (Wimmer, 2014, p. 188). If war's consequence was welfarism, welfarism itself accompanied consequentialism as a utilitarian component. Sen (1999, p. 62) illuminates welfare's utilitarian foundation, consequentialist ethics and the teleological links. Sen recalls the Benthamian definition of utility as 'pleasure, or happiness, or satisfaction'. These mental states are judging criteria for a state of affairs. Rights, duties and the kind become irrelevant information. When welfarism and consequentialism combine, teleology shines. For Sen, choice is judged by utilities generated. An 'action is judged by the consequent state of affairs (because of consequentialism), and the consequent state of affairs is judged by utilities in that state (because of welfarism)' (Sen, 1999, p. 59).

The two world wars hastened expansion of nutrition and health care policies in Britain. The first war saw exceptional development of 'sharing' attitudes and policies. During the second war a sharing psychology in 'beleaguered' Britain radicalized public arrangements for distributing food and health care. Nutritional studies reveal decline in undernourishment and disappearance of extreme undernourishment.

In India, imports of industrial British goods displaced craftsmen, for instance. This disrupted economic self-sufficiency. The colonial administration did not react. The English, among the first nations to 'install a system of poor relief at home, perceived the problem of the rural poor as a result of their backwardness, untouchability and of "Indianness" itself' (Ehmke, 2012, p. 86). The administration's efforts to establish social security hardly went beyond, to quote Loewe (2009), the 'small segment of the population whose contentment was particularly important for the colonial powers to stabilize their rule' (Ehmke, 2012, p. 86).

The effects of World Wars II and I were similar. Excess demand and supply constraints developed. But in 1939, industrial incomes increased and industrial diversification was wider. But India became the war's eastern front and 'Bengal was now a theatre for war'. The British requisitioned rice in anticipation of a long war. Harvests failed and food prices rose in eastern India. Millions perished (Roy, 2011).

The 'moral salience' of World War II and a later 'global consciousness' in resolving pre-war inequalities and asserting democratic rights of self-determination led to decolonization and development through modernization. Democratic commitment and sensibilities underlay the idea of development. It involved the removal of inequalities between and within nations through production and attaining affluence. The ecological constraints to economic growth remained extraneous during the development decades that followed (Bell, 1998; Guha, 2000). The ecological consequences of the modernist project are evident in high rates of deforestation. In the post-war development decades, rainforests and other natural habitats in the equatorial and third world tropics were cleared (Bell, 1998; Mazur, 1999).

The social consequences of deforestation are manifest in forest access and control conflict. Involving forest dwelling and dependent communities, conflicts have assumed varied forms of local and community movements. In trying to retain control over resources threatened by statization or the progress of the market, such conflicts reveal an ecological content (Guha and Martinez-Alier, 1998). Production-based conflicts are also complemented by protective variants. Protected Area regimes after initially finding a colonial constituency in Africa have contemporarily, since the post-war period, spread in the form of national and wildlife parks and biosphere reserves in developing tropics.

India denuded and converted its forests for post-independence development. Development was typically a process of industrialization and urbanization. The political economy of post-colonial development in India was characterized by the wresting of concessions by a capitalist class. It appropriated nationalism despite the theoretical primacy accorded to the peasant. The choice of an industrial model of economic growth was conditioned by capitalist merchants and industrialists, the technical and administrative bureaucracy and rich farmers. The state granted liberal access to forest resources required for industrialization. A process in which it had largely invested. Resource extraction was regulated through bureaucratic rules and regulations. They facilitated control over the extractive process (Gadgil and Guha, 1992).

The passage of Indian forests through contemporary industrial stages also reflect deforestation phases. Four million hectares of forests were felled between 1951 and 1976. Over the next 15 years, 22.7 million hectares were cleared (Baviskar, 1995). Commercial forestry has intensified conflicts between commercial-industrial interests it serves and rural interests it thereby denies. Conflicts over shifting agriculture have increased in regions of northern India. Conflicts between settled cultivators and the state have however persisted in the Himalayas and the Western Ghats. Conflicts have attenuated only in regions where forests have lost significance (Baviskar, 1995; Sundar, 1997). In debates perceivably on 'environmental' issues, for example, Chipko and the Silent Valley, there was a sharp division between those who supported 'development' over 'environment' and between those who argued for 'environment' over 'development'. The 'development' camp sought to accelerate industrial economic growth for increasing human welfare. The 'environment' camp sought to preserve ecosystems and endangered species and maintain 'overall ecological balance' (Baviskar, 1997, p. 196).

Ecological Marxists sought to falsify the development versus environment dichotomy. India's development policies failed because 'they had not fundamentally changed the patterns of control over the means of production'. Means of production concentrated 'in the hands of the state bureaucracy, the industrial elite and the rich peasants' (Baviskar, 1997, p. 197). This 'skewed technological choices, production decisions and income distribution'. Development based on 'uncontrolled industrialization was bound to fail' (Baviskar, 1997, p. 197). It destroyed natural resource bases on which material prosperity is founded. It also destroyed livelihood sources of the poor (Baviskar, 1997).

Livelihood studies and projects have proliferated since the 1980s. Later in the decade, sustainable livelihoods gained popularity after the Bruntland report, 1987. Livelihoods entailed an element of agency. In a conservation context it was about purposive human strategies in accessing and accumulating natural resources, actions that cope with and adapt to shocks and stressors, be they forest regulations or animal raids. But in 2005, a discourse that gave agency to ecosystems emerged – ecosystem-serviced human well-being. The services approach emerged when well-being of forest communities was central to the livelihoods approach. Rounding-up reviews of poverty, sustainable livelihoods and biodiversity conservation-based well-being approaches sets good context to discuss the services framework.

Poverty, professional economics and local realities

In the 1980s and 1990s, 'anti-poverty' was national and international rhetoric. But economist-dominated 'reductionist and standardized' professional realities prevailed over the 'local, complex, diverse and dynamic' rural realities of the south (Chambers, 1995, p. 173). Poverty was reduced to income-poverty and employment to jobs. But income-poverty was only an aspect of deprivation. As participatory rural appraisals made evident, people experienced multiple disadvantages including 'social inferiority, isolation, physical weakness, vulnerability, seasonal deprivation, powerlessness and humiliation' (Chambers, 1995, p. 173). Income-poverty's 'pre-eminence' has to do with the development discourse's domination by economists. Poverty was a concept conceived in predominantly northern industrial and urban contexts. Here cash income and cash-based consumption indicated economic status. Economists speculated that the poor preoccupy with income and consumption.

In world development and sustainable poverty reduction thinking of the 1990s, income served as end. Education and healthcare were justified as means. Education and health were not considered as justified ends themselves or even as self-respect and capabilities enhancing aspects. They were hard to measure. 'Deprived people are often thought of as being uniform' (Chambers, 1995, p. 188). 'But, if anything, there is more diversity among the poor than among the non-poor'. Self-respect, with freedom from dependence, is a 'dimension most overlooked and undervalued by professionals' (Chambers, 1995, p. 190). Humiliation and self-respect do not lend themselves to measurement. Their living is improvized and sustained through their livelihood capabilities, and through tangible and intangible assets. However, capabilities, tangible and intangible assets are not separate livelihoods requirements (Chambers, 1995).

So, did sustainable livelihood approaches indeed go beyond income and include other deprivations? How does capabilities figure in these approaches?

Sustainable livelihoods: being inclusive about indicators and outcomes

Scoones's (1998) much-cited Institute for Development Studies (IDS) working paper proposed a 'capitals' and institutions-based framework for sustainable rural livelihoods. He draws on Chambers in his capabilities inclusive definition of livelihoods. A livelihood comprises capabilities, and social and material resources. A sustainable livelihood copes with and recovers from stresses and shocks. It maintains or 'enhances its capabilities and assets without undermining the natural resource base' (Scoones, 1998, p. 5).

For Scoones, Sen's notion of capability (1984, 1987) enhances the definitional scope of livelihoods. Capabilities are what people can do or be with their entitlements. It goes beyond material concerns of food intake or income. This well-being approach to livelihoods allows 'people themselves to define the criteria which are important' (Scoones, 1998, p. 6).

Sustainable livelihoods brought together different development debate strands. It gained traction under the umbrella of poverty alleviation and biodiversity conservation. But it seems that trade-offs were not made explicit in studies that reduced well-being to economic and political aspects. Such studies failed to set context and were hardly comparative.

Biodiversity conservation, poverty and well-being: understanding and considering trade-offs

Agrawal and Redford (2006) provide a pithy review of poverty alleviation and biodiversity conservation. They studied poverty and biodiversity programmes in ten southern countries. These interventions assume three types, namely, community-based wildlife management, extractive reserves, and ecotourism. They compare these interventions to conceptual literature around poverty and biodiversity. They concluded these interventions 'involve simplified understandings of poverty and biodiversity' (Agrawal and Redford, 2006, p. iv).

Of 47 studies the authors reviewed, only two remained comparative and the rest were single case studies. These empirical studies operationalized poverty and biodiversity differently from theoretical discussion of them. In theory, poverty is multi-dimensional and biodiversity has multiple attributes and components. Achievements on one dimension or attribute cannot be equated with or translated into another. But these empirical studies were 'indifferent to these critical differentiations' (Agrawal and Redford, 2006, p. 29). They focused on economic, and less frequently, political aspects of poverty. There was rare attention to health, educational and cultural dimensions of poverty. Descriptive statements of poverty impacts abound. Similarly, biodiversity measures focused on species diversity, 'usually simple measures of presence/absence of selected species e.g. butterflies, trees or relative abundance of select species e.g. lowland gorillas' (Agrawal and Redford, 2006, p. 29). The authors find that these studies limit generalizability and suggest broadening of the field. There is a need to first understand 'how changes in biodiversity and poverty are related' (Agrawal and Redford, 2006, p. 33). It is also necessary to use methods that generate 'better and deeper understanding of these changes and their trade-offs' (Agrawal and Redford, 2006, p. 33).

So there cannot be comprehensive conservation and well-being win-wins. Trade-offs need to be explicit and acknowledged. And so, how have studies of interventions conceptually and methodologically coped with this critique over ten years since Agrawal and Redford's review? The following is a review of well-being studies in environmental contexts. They do not engage with biodiversity or ecosystem changes and complexities but do well in complicating well-being changes in terms of trade-offs between singular and multi-dimensional definitions. They also consider methodological trade-offs concerning qualitative and quantitative measurements.

Environmental interventions and well-being: practical trade-offs

Fry et al. (2017) consider conceptual and operational trade-offs that policy makers and practitioners face when applying well-being within the context of environmental interventions. Reviewing well-being in terms of a conceptual trade-off, namely its operationalization in single and multi-dimensional ways, the authors trace objective, subjective and hybrid definitions. While objective definitions do not 'so much [look] at well-being itself as at the factors that influence well-being', subjective approaches consider how 'external, objective factors are translated into internal experiences of well-being', whether satisfaction, pleasure or pain. Hybrid definitions that emphasize quality of life, blending objective and subjective approaches 'transcends both by recognizing the way each is socially constructed'. Sen is acknowledged as having pioneered this 'mixed approach by emphasizing that people are likely to subjectively experience their objective deprivation very differently' (Fry et al., 2017, pp. 69–70).

Although the 'hybridized definition promotes a holistic approach to monitoring the impacts of interventions', such inclusiveness will make well-being tough to operationalize. Policy makers and practitioners typically face measuring trade-offs (Fry et al., 2017). While singular definitions are useful in monitoring intervention impacts, they are narrow and describe well-being changes only partially. Multi-dimensional definitions render change monitoring and measuring complex. In reconciling this trade-off, policy makers and practitioners may choose appropriate indicators. Quantitative approaches, underpinning externally valid interventions, generate numerical data. Qualitative approaches generate meaningful well-being data useful in internally valid or contextual interventions. But policy makers prefer quantitative data.

Agrawal and Redford do not mention capabilities in their review of biodiversity and poverty programmes. These predominantly used income and institutions as indicators and outcome. But Fry et al. in discussing the drawback of subjective well-being definitions cite Keyes and Waterman's (2003) suggestion to use hybrid definitions. This includes 'internal states as well as external objective capabilities' or positive functioning and successful accomplishment. Further in their discussion of well-being definitions and various trade-offs where the authors advocate for an internally validated monitoring of well-being, capabilities is also hinted as an approach that can form the basis of, or suit, their advocacy of qualitative and contextual well-being. They cite Deaton (2008) about Sen's widely accepted view that human well-being depends on a range of functions and capabilities enabling people to lead good lives. These capabilities cannot be aggregated into a single summary measure (Fry et al., 2017).

There is the need to integrate qualitative approaches like capabilities and human narratives with dominating quantitative measures. Besides policy fondness for quantitative indicators and outcomes e.g. income, natural sciences have ensured the domination of externally validated quantitative monitoring of well-being. Ecologists have commonly been involved in monitoring intervention impacts on well-being, 'an area in which they may not have sufficient experience or training'. Natural scientists 'tend towards concise, quantitative results to evaluate the impact of their projects and demonstrate success to donors and colleagues'. Like Chambers' plea that people's own articulation of their needs be listened to, Fry et al. recommend that policy makers and practitioners be made aware of the need to give local well-being perspectives more attention. This increases the local relevance of indicators and enhances stakeholder equity (Fry et al., 2017).

How, if at all, did the ecosystem services framework engage with local articulations of livelihood capabilities? What about links between changes in biodiversity and well-being? And what about the need for qualitative accounts of locally meaningful well-being?

Ecosystem services and human well-being

In February 2014, the Government of India approved the National Mission for Sustaining the Himalayan Ecosystem (NMSHE), under the National Action Plan on Climate Change (NAPCC). The Mission sought to assess the health of Himalayan ecosystems for policy formulation. These policy measures were envisioned to sustain ecological resilience and ensure continued provisioning of key ecosystem services. The Millennium Ecosystem Assessment (MEA) report of 2005 mainstreamed the ecosystem services concept. Like natural historians and ecologists who developed species taxonomy, the MEA team introduced taxonomies of function.

> Thus, a forest's function of storing and letting-off water became a 'regulating service'. Producing edible leaf and fruit became a 'provisioning service'. Forests and grasslands provisioned spatial context for recreation, aesthetics, inspiration and spirituality, which were 'cultural services'. Such services are benefits that humans experience.
>
> (Krishnan et al., 2017, p. 37)

Ecosystems, traditionally considered to provide economic goods of tangible value like food and fodder, are contemporarily considered to also provide services. This includes flood regulation and beauty creation. In the process of its valorization variedly as an economic instrument, conservation medium and policy fix, the idea of ecosystem services has ignored human roles and capabilities in creating and receiving services and adapting to changes in services.

The ecosystem services framework has been subject to much critique that bear well-being implications. As like any other framework that has become popular and has persisted, ecosystem services has also amassed enough policy and academic heft so as to attract critical reviews. Let us discuss two such, namely Lele et al. (2013) and Schroter et al. (2014). Lele et al. discuss well-being implications of ecosystem services in more detail, but do not engage with capabilities or for that instance any other well-being approaches. Schroter et al., in classifying 'recurring critiques' mention capabilities as a way out of one particular critique, namely of ecosystem services valuation (Schroter et al., 2014, p. 514).

Citing Norgaard (2010), Lele et al. point to ecosystem services originally serving as a metaphor. It became basis for measuring, assessing and valuing social dependence on nature. While the MEA framework does not promote an economic valuation of well-being, 'the vast majority of social scientists associated with the MEA are economists' (Lele et al., 2013, p. 344). Most ecosystem services assessments are economistic in practice or intent. However the 'the idea that life on Earth is fundamental to human well-being', and hence not to be compromised, is being accepted for the first time in wider circles (Lele et al., 2013, p. 345).

While ecologists agree that ecological functions are not services, terming ecological processes confusingly as supporting or intermediate services is a tendency of attributing agency or purpose to ecosystems. Yet, there is no evidence that supports such 'grand design' of nature for human well-being (Lele et al., 2013, p. 347).

Human labour also plays a role in obtaining benefits from ecosystem processes. Water-flows for instance are useful if 'lifted, diverted or stored'

(Lele et al., 2013, p. 351). Natural capital, whose abundance is to also increase ES, which further increases well-being, also results in disservices. Large carnivore depredations are an apt example.

Unlike Lele et al., whose review of ES was more exhaustive, Schroter et al. synthesize recurring ES critiques and counter-arguments. These include contestations that ecosystem services strategies divert attention from biodiversity and also that links between services and biodiversity components are weak. This is countered by arguments that the ecosystem services frameworks acknowledge biodiversity-services overlaps, for instance, through biodiversity roles in cultural services. Also, the empirical evidence on high biodiversity is necessary for provisioning and regulating services. Capabilities appears in counters against contestations that ecosystem services is framed in economic terms and that assessments involve valuations. The counter-argument is that services does not involve only monetary valuation and that it 'can be used to assess human well-being according to the capability approach, which deals with people's freedom to live a good life' (Schroter et al., 2014). Finally, contestations of vagueness of concept are countered by arguments that definitions improve and flexibility inspires trans-disciplinarity.

The suitability and scope of capabilities

These reviews suggest suitability and scope for the capabilities approach in assessing well-being in ecological contexts. In the 1980s and early 1990s, poverty professionals had prevailed over local communities. They privileged measurable and singular well-being indicators like income. Education and health were not considered as meaningful ends by themselves, nor as meaningful means that enhanced dignity and capabilities. Livelihoods, an approach that can bridge expert and lived realities, accounts for besides tangible and intangible assets, people's capabilities in improvising and sustaining their lives. A livelihood is sustainable when it adapts to shocks and stress, and its capabilities are maintained or enhanced without affecting natural resources. Income generation need not be the sole significant sustainability indicator. The livelihoods concept and definition stands enhanced by the inclusion of quality-of-life philosophies including capabilities. It encompasses more than food or income and is about how people can function with their entitlements. In well-being monitoring that is locally sensitive and meaningful, capabilities can constitute a qualitative assessment. Ecosystem services assessments need not have to involve valuation but also be assessed in capability terms.

Combining capabilities and ecosystem services

Ecosystems are dynamic. They change in response to biotic and abiotic factors. Human actions, intended and unintended, are regularly responsible for such biotic and abiotic changes, whether by introducing an exotic species or increasing greenhouse gas emissions. Human organization is also impacted by ecosystemic changes. Also, like ecosystems, human organization is dynamic. It adapts to ecological change, even as it differs in its abilities to adapt. The idea of trade-off that is central to ecosystem service configurations becomes conceptually and rhetorically pertinent here. Trade-offs are not just about advantages and disadvantages that arise in, for instance, the privileging of a regulatory service over a provisioning service, but also about the advantages and disadvantages that accrue from the very privileging of the concept of ecosystem services. To claim that the ecosystem is a sentient agent that provides and regulates intentionally is to ignore the role of (structured) human agency in co-creating such services in the first instance. Addressing this ontological trade-off will, among other things, thaw sociological and anthropological indifference to the concept.

From an environmental sociological consideration, ecosystem services is a discourse requiring critical analysis for silencing human labour and technology in shaping services that an ecosystem ostensibly provides or regulates. Also, concomitantly, the bestowal of agency to the ecosystem as a 'service provider and regulator'. A rational agency whose services have 'exchange' or monetized value. But, discourse is a critical field of enquiry that social constructionists, those who investigate culture's role in how environmental ideas, concepts or problems are produced, prefer, sometimes exclusively. Such an approach would not facilitate a long collaboration over ecosystem services with ecologists. A more 'realist' engagement that does not totally eschew notions of discourse would be to engage with the idea of ecosystem services as a potentially useful way of reconfiguring society-nature relations and how human-human relations effects and conditions the well-being benefits and burdens that emerge from this relationship. But, there must remain within such a realist approach scope to assess how communities themselves consider and experience the ecosystem, and their inequitable roles in co-producing its services.

Some questions immediately arise here. These include – what are these service creating and receiving capacities? What capacities are implicated in responding to changes in services? Are there differences in human capacities? Here lies the rationale for the formulation of an ecosystem and well-being assessment agenda in a manner that is being attempted here. It is anchored as it is to the conceptually more sophisticated synonym of capacities, namely, capabilities. It is agenda that is avowedly more qualitative, political and philosophical than regular ecological economic assessments, those that aggregate economic well-being in narrowly monetary terms.

The quality of community lives and capacities to ensure dignified quality are well-being matters central to (social) science and policy debates surrounding ecosystems and the environment. Poor people experience chronic and acute environmental stressors. Understanding the differentiated capacities of individuals, households and communities to respond to environmental stress has been strongly influenced by Amartya Sen's (1999) capabilities approach (Brown and Westaway, 2011).

For Martha Nussbaum (2011) who elaborated his theory in ways that also made it environmentally relevant, Sen identifies capabilities as the space for comparing life quality assessments. The assessment is fundamentally about what each person is able to 'do' and 'be'. The impetus is on freedom and choice. A society should promote among its people opportunities and substantial freedoms. However, people have the choice to exercise or otherwise such freedoms. Capabilities, then, are substantial freedoms to choose and act. Being and doing are about functioning. Sen's formulation of a person's capabilities entails feasible 'alternative combinations of functionings' (Sen, 1999, p. 75). To function is not just an internal or mental-physical capacity, but also a capacity enabled by socio-political and economic environments.

How does Sen himself link development policy, freedom and capabilities? Political freedoms, economic and social opportunities, transparency and security are rights that advance a citizen's capabilities. Freedoms of individuals are basic development building blocks. Development need pay attention to expanding capabilities of people to 'lead the kind of lives they value and have reason to value' (Sen, 1999, p. 226). Public policy can enhance capabilities. Capabilities can enhance public policy in a participative manner. Evaluation and effectiveness are two distinct reasons for why individual freedoms are crucial in development. Freedom to pursue things one has reason to value is 'significant in itself for the person's overall freedom' (Sen, 1999, p. 18). It is also 'important in fostering the person's opportunity to have valuable outcomes'. These are relevant in evaluating 'freedom of the members of the society and thus crucial to the assessment of the society's development'. The second reason for substantive freedom's significance is in determining 'individual initiative and social effectiveness. Greater freedom enhances the ability of people to help themselves and also to influence the world' (Sen, 1999, p. 18).

But to evaluate capabilities, we must determine which functionings matter and how much. We need a valuation procedure. Listing of flourishing constituents is one way. Nussbaum does this. Sen rejects this approach. It denies the relevance of the values people may come to have and the role of democracy. People must be the legitimate source of decisions about reasoned and valued living. Sen proposes social choice exercise involving public reasoning and democratic procedures of decision-making (Plato.stanford.edu, 2019).

Listing central capabilities

Nussbaum (2011), unlike Sen, goes on to list ten capabilities and suggests thresholds. She founds her approach upon notions of dignity and social justice. It asks what a life worthy of dignity requires. At bare minimum, an ample threshold of ten capabilities is required. She subscribes to a wide understanding of the government's role as enabling people to pursue a 'dignified and minimally flourishing life'. A 'decent political order' need ensure at least a threshold level of these ten capabilities (Nussbaum, 2011, p. 33).

While the eighth capability explicitly pertains to species and nature, the remaining capabilities possess potential linkages with conservation and natural resource questions.

1 Life: Capability to live a worthy life of normal length.
2 Bodily health: Capability to have an adequately nourished and good health.
3 Bodily integrity: Freedom of mobility and security against violence.
4 Senses, imagination and thought: Capability to imagine, think and reason in human and informed ways, cultivated by education, literacy, basic mathematical and scientific training.
5 Emotions: Capability to possess attachment to things and people.
6 Practical reason: Capabilities to form conceptions of good and critically plan one's life.
7 Affiliation: Capability to live with and show concern for other humans. Capability to socially interact with others and imagine their situations.
8 Other species: Capability to live with concern for and in relation with animals, plants and nature.
9 Play: Recreational and play capabilities.
10 Control over one's environment: Political – capability to politically participate; and Material – capability to equally hold landed property and seek equal employment.

(Nussbaum, 2011, pp. 33–34)

The 'other species' capability may appear an idealist and instrumental one. At least inasmuch as concern for, and relations with, animals, plants and nature are considered ethical and empathetic ends that satisfice humans. However, Nussbaum postulates potential positions that could be assumed variedly, beyond the fundamental consensus that only human capabilities count. These positions are:

1 Human capabilities are ends in themselves, though non-human capabilities may be instrumentally valuable.
2 Non-human beings' role in relationships with humans not just means but with intrinsic value to the relationship.
3 Capabilities of all sentient beings as ends in themselves and, requiring threshold attainment.
4 Capabilities of flora and fauna as ends, but as individual entities and not as ecosystem constituents.
5 Ecosystem capabilities count as ends in themselves.

Nussbaum anticipates a vigorous debate among capability analysts about assuming any of these five positions (Nussbaum, 2011). If such a debate were to occur among the numerous ecosystem services advocates, they would concur either with positions one, two or five, depending on the strength of their convictions over the role of ecosystems, and further of flora and fauna within them. Inasmuch as position five is concerned, it needs noting that ecosystem services advocates pitch such services as a crucial means to human well-being, whatever be the merits of intrinsic value, and so for the purpose, services constituting an end by themselves. By implication, position two gains significance.

Of the various positions assumed, for Nussbaum, environmental quality plays a role in the capabilities approach. She states the non-requirement of needing to move beyond position one, to argue for the criticality of the ecosystems health for human well-being, especially if well-being entails commitments to future generations (Nussbaum, 2011). This position is synchronous with the sustainability philosophy and the MEA's goal of enhancing ecosystem services' contribution to human well-being without compromising their long-term productivity. The way to include the interests of future generations is significant if the capabilities approach is to emerge a serious player in the environmental realm (Nussbaum, 2011).

Environment as 'meta-capability' or 'sustainable ecological capacity'

Breena Holland's (2008a, 2008b) work on ecology being a 'meta-capability' and the need to establish 'capability ceilings' in terms of ecological limits to justice are for Nussbaum important recent works. They argue how the capabilities approach offers advantages over other approaches favoured by environmental economics mainly because the wide effects of environmental negativities are disaggregated over different aspects of human lives. Nussbaum considers Holland's approach as instrumental and thus anthropocentric. This, not because the latter believes in such a stance, but because the anthropocentric premise is an uncontroversial one that most public presumably accept (Nussbaum, 2011, p. 164).

Holland's instrumentality entails treating 'environmental entitlements as a matter of basic justice'. She treats 'certain environmental conditions as instrumental to human capabilities' (Holland, 2008, p. 319). This is in the same way that Nussbaum 'treats material things such as shelter, nourishment and property as instrumental to human capabilities'. Because 'environmental conditions are necessary for producing and sustaining material value', Holland seeks to 'establish these environmental conditions as an independent "meta-capability"' (Holland, 2008, p. 320).

By environmental conditions, Holland means environmental resources and services. But she does not specifically mention ecosystem services. However, both services and ecosystems are considered crucial to capabilities. For instance, Holland's meta-capability, by which she means 'Sustainable Ecological Capacity', implies 'an individual's opportunity set' and not just 'a property of a nonhuman system'. The two primary opportunities at stake in protecting Sustainable Ecological Capacity as an individual capability are:

1 The opportunity to individually exercise any of Nussbaum's human capabilities, which require ecological systems to maintain the conditions of life; and
2 Individual opportunities to 'exercise capabilities in relation to the environmental resources and services that the current generation enjoys'.

(Holland, 2008, p. 324)

The implications for ecosystem services are manifest in each of the two opportunity sets just mentioned. In the first, capability exercise opportunities are dependent on ecosystems maintaining life conditions. In the second, environmental (read ecosystem) services whose opportunities of use for future generations need to be at par with the present generations.

However, Holland's 'meta-capabalizing' of ecosystems does not amount to bestowal of agency to ecosystems. Ecosystems are not treated as 'ontological entities'. They are not conceptualized as 'having opportunity sets from which they can make choices'. Rather, Holland claims that given the extent of human capability dependence on the environment, the entitlements it provides needs treatment as a basic justice matter (Holland, 2008, p. 320). Thus, ecosystems or its services, when considered a meta-capability, perhaps refer more to what humans can 'do' (use) with water, soil or wood or 'be' with regard to stewardship of such resources and services, rather than refer to an ecosystem as a sentient, decisive and choosing entity (Holland, 2008).

As Holland notes, Nussbaum defined her list of central human capabilities abstractly so that they could be nationally and locally contextualized. In the following section we set context – ecological, economic and cultural. By cultural we mean not just ethnic identities of Darjeeling's people, but also the disciplinary identities of the research team behind this book.

The Eastern Himalayas: ecology and economy

The Eastern Himalayan region comprises the countries of Bhutan, Nepal and the Indian Eastern Himalayan states. This region is one of the 34 global hotspots of biodiversity (Mittermeier et al., 2004). Multiple biogeographic origins, altitudinal variations, climatic variability and topographic complexities have made the region extremely rich in floral and faunal diversity; 10,000 plant species, 240 mammal species and 750 bird species are known to occur across this region's rugged and remote landscapes. Besides tremendous richness of species, many endemic, the region fosters a diversity of habitats, from the flood plain savannas to Himalayan alpine meadows. They are a source of ecosystem services that support livelihoods and economies of local as well as downstream communities.

Biodiversity is complemented by cultural diversity, with a large number of indigenous and immigrant ethnic groups, like *Rais, Sherpas, Lamas* and

Tibetans inhabiting the Eastern Himalayas. The material and cultural well-being of these communities are variedly dependent on the forests and pastures. These ecosystems provide food, fibre, fodder, fuelwood, medicinal plants; are habitat for wild pollinators; and regulate climate, water and carbon. As many other ecosystems, the Himalayas also provide aesthetic and serene space for recreational and spiritual pursuit. Importantly, the Eastern Himalayas, like its western counterpart, discharges water to millions living downstream.

The last three decades have seen the Eastern Himalayan region undergo large-scale changes that have impacted biodiversity and livelihoods. Conversion of forests into other land uses has been conspicuous due to rapid economic development and expanding agriculture. Developmental activities like road building and mining have diminished and degraded natural habitats. The many hydropower projects commissioned have reduced forest cover, eroded soil and degraded riverine ecosystems. The long-term livelihood impacts are to be studied (Khaling and Pradhan, 2017).

Darjeeling

Sikkim and Darjeeling are a contiguous Eastern Himalayan geography punctuated only by political and administrative boundary. A biodiversity hotspot, the region bears further significance as Bengal's water catchment. Climate change and land use change renders both this status and the significance vulnerable. Discounting for Darjeeling's colonial tea plantations, Sikkim and Darjeeling had remained largely rural as a culture, geography and economy for much of the 19th and 20th centuries. Tourism, earlier an urban phenomenon entailing summer visits to Darjeeling and Gangtok, is now an all-penetrative phenomenon that economically links the region to the rest of India and the world. Forest, pasture and village all now need to adapt to unprecedented visitor footfall (Krishnan et al., 2017).

From 300 metres (m) at base, Darjeeling's elevation reaches an altitude of 3,660 m at Sandakphu on the Singalila ridge. Darjeeling's altitudinal gradient is responsible for its floral and faunal distinction, with high altitudinal range ensuring climatic variability. This is 'normal climate change so to speak, where temperature, precipitation and humidity vary quickly across geography and foster diverse vegetation ranging from the tropical to temperate to sub-alpine and alpine' (Krishnan et al., 2017, p. 38).

In the Darjeeling Himalayas, forests have been converted in large scales into tea gardens during the colonial period and to Cinchona plantations at smaller scales. Forests are not contiguous. There is a certain patchiness to the landscape. West Bengal's famed Joint Forest Management (JFM) programme has not translated into any robust community-based forest management institutions in its northern hilly reaches.

After four years of protests for statehood, in 2011 the Gorkhaland Territorial Administration (GTA) replaced the Darjeeling Gorkha Hill Council.

This Council administered the hills for 23 years since 1988. The GTA has jurisdiction over sectors including reserve forests, agricultural lands, water resources and tourism. The Singalila and Senchal Protected Areas are administered under the Wildlife Act, 1972.

The ecological and well-being assessments

The MEA attempts an interdisciplinary stance when it states that an assessment's usefulness to different stakeholders will also depend on the composition of the scientific community that conducts it. A team of ecologists and an environmental sociologist, we conducted ecological and well-being assessments of forest ecosystems and forest-dependent farmer households between 2015 and 2017 in Darjeeling. We assessed the relationship between ecosystems and human capabilities as relevant in the regional-local context of east Himalayan forest conservation and development. The assessments, conducted in seven villages near Singalila National Park and Senchal Wildlife Sanctuary, involved rigorous biodiversity and water assessments, as well as capabilities-based well-being assessments through household surveys (see Figure 1.1).

Senchal Wildlife Sanctuary was established by the British in 1915. It is a 38.60 square kilometre forest, known for the Asiatic bear population.

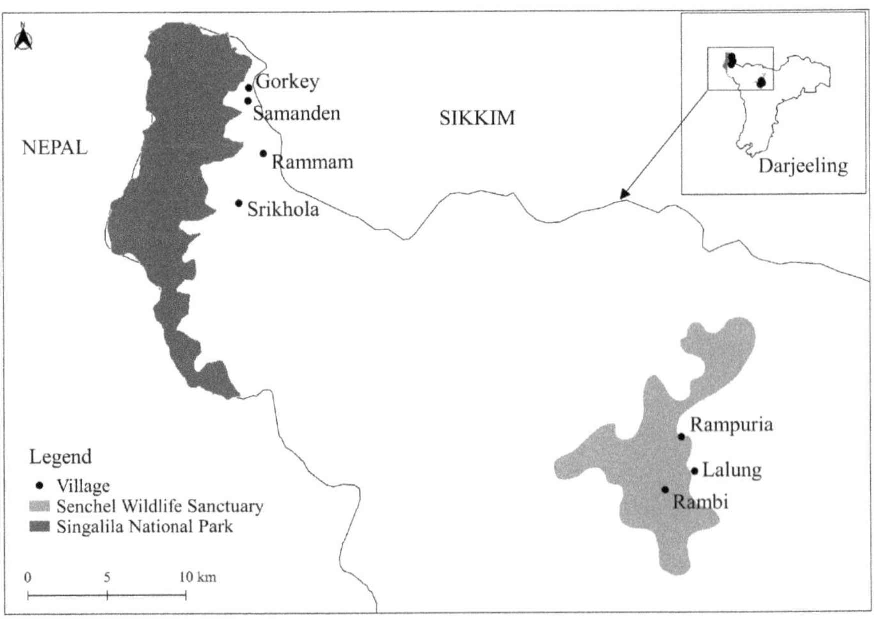

Figure 1.1 Darjeeling: the study areas

Source: Generated by Abhishek Samrat.

Cinchona plantations have been raised near the sanctuary. We chose three villages in and around Senchal. These include Lalung, Rambi and Rampuria. Singalila was declared a wildlife sanctuary in 1986 and was made an Indian National Park in 1992. Here, we chose four villages to assess, namely, Gorkhey, Samanden, Rammam and Srikhola.

Each of these villages appear in different combinations in this volume's chapters. This in keeping with the comparative scope and spirit of life quality assessments that the capabilities approach encourages. Chapter six is an exception, where a single hamlet is evaluated for what it values and of what it is deprived.

Village profiles

Of the villages near the Singalila National Park, Gorkhey, Samanden and Rammam are forest villages. The latter two fall under the Forest Department's territorial jurisdiction. The former comes under the wildlife department's jurisdiction. These four villages have no land title. Among the Senchal villages, Rampuria and Srikhola are revenue villages and households have land title. Lalung is a Cinchona plantation settlement turned village with no land title. Rambi is a forest village under the wildlife department jurisdiction. Detailed profiles of these villages follow.

Gorkhey Forest Village

Gorkhey Forest Village (Table 1.1) is situated in the Singalila North Range under the Darjeeling-Sadar sub-division of Darjeeling district, West Bengal. The village is situated between the latitude of 27°11.329'E to 27°11.225'E and longitude of 88°04.367'N to 88°04.397'N. It is at a slope in the south-east direction. The village falls under the wildlife jurisdiction of the Singalila National Park. The Rammam River flows on the eastern side of the village. From the western side, the Gorkhey River enters the village and joins the Rammam River and the combined river flows as the Rammam River.

Table 1.1 Village profile: Gorkhey Forest Village

Gram Panchayat	Sirikhola-Daragaon
Block	Darjeeling Pulbazar
Sub-division	Darjeeling-Sadar
District	Darjeeling
Forest Range	Singalila North Range
Beat	Gorkhey
Number of Households	32
Major Religion	Buddhism and Hinduism
Altitude	2,400–2,600 m

Source: Fieldwork.

The village is situated at the border of West Bengal and Sikkim. The Rammam River separates the two states. The nearest village is the Samanden Forest Village, which is situated at the southwestern side of the Gorkhey village. Samanden falls under the territorial jurisdiction. The nearest village of Sikkim is Bharang, which is 5 kilometres away. Bharang has road connectivity in all weather. One shared vehicle plies to Jorethang town of West Sikkim in the morning and returns to Bharang in the evening. It takes almost three hours to go to Jorethang from Bharang. Gorkhey village has no road network. People use foot tracks to go either to Bharang or to Shepi village at a distance of 24 kilometres in West Bengal to access motor vehicles. People either walk or ride on horses to reach nearby villages.

There is one primary school at Samanden. Most children of Gorkhey and Samanden study there. For upper primary, secondary and higher secondary education, people send their children to different schools in West Sikkim and Darjeeling.

Gorkhey does not have electricity connection owing to the objection from the Wildlife and Forest Departments for establishing electric posts within the protected area. Some of the households have installed solar photo-voltaic plates for generating electricity for tube lights and charging mobile phones. The Forest Department is reluctant to provide the no-objection certificate for an electricity connection line since the wire has to be brought through the forest.

The village is mainly inhabited by Buddhist and Hindu population. People speak in Nepali. The main occupation of the people is agriculture. Every household has a small amount of agricultural land where they cultivate different crops like maize and other vegetables. They generally cultivate two crops a year. The temperature during the winter falls below 5 degrees Celsius. Hence, they cannot do farming activity during that time. Other occupations are livestock rearing, tourism, working as labourers in Darjeeling and Sikkim, etc. The people rear cows, goats, pigs and poultry birds. There are three tourist lodges in Gorkhey village that earn tourist and trekker income. Villagers also earn cash income through the Mahatma Gandhi National Rural Employment Guarantee Scheme (MGNREGS), either implemented by the Forest Department or the *Panchayat* Department. The Forest Department also provides wage labour for raising plantations, construction of foot-tracks and drains.

People take their agricultural and livestock products to the markets of Bharang or Rimbik either by horse or on their back. People buy different things from the market like rice, wheat, vegetables, meat, utensils, clothes and other necessary items.

The village has a *Gram Samaj* of which all adults of the village are members. *Gram Samaj* has an informal governing body. Its main function is to help any of the households during crises. The main formal institution functioning in the village is the Eco-Development Committee which takes care of the village development and conservation of the forest.

Samanden Forest Village

Samanden Forest Village (Table 1.2) is adjacent to the Gorkhey Forest Village and quite similar to the latter. The distance between these two villages is less than 1 kilometre. Samanden lies to the west of Gorkhey. However, there is a difference in the topography of these two villages. Unlike Gorkhey Forest Village, Samanden is situated on the top of one of the hills and is more or less flat. It is one of the forest villages of the reserve forest and there is one Forest Protection Committee which looks after village development and conservation-related issues.

Rammam Forest Village

Rammam Forest Village (Table 1.3) is situated in the Singalila North Range under the Darjeeling-Sadar sub-division of Darjeeling district, West Bengal. It falls under the territorial jurisdiction of the Forest Department. The village is situated at a latitude of 27°09'42.09"N to 27°09'6.78"N and longitude of 88°04'30.85"E to 88°05'13.44"E. The slope of the village is towards the southeast direction. The Rammam River flows on the eastern side of the village and separates Rammam from Sikkim Province. The village opposite to the river on the Sikkim side is Piuretar.

Table 1.2 Village profile: Samanden Forest Village

Gram Panchayat	Srikhola-Daragaon
Block	Darjeeling Pulbazar
Sub-division	Darjeeling-Sadar
District	Darjeeling
Forest Range	Singalila North Range
Beat	Samanden
Number of Households	28
Major Religion	Buddhism and Hinduism
Altitude	2,600 mm

Source: Fieldwork.

Table 1.3 Village profile: Rammam Forest Village

Gram Panchayat	Sirikhola-Daragaon
Block	Darjeeling Pulbazar
Sub-division	Darjeeling-Sadar
District	Darjeeling
Forest Range	Singalila North Range
Beat	Rammam
Number of Households	52
Major Religion	Buddhism, Hinduism, Christianity
Altitude	1,975–2,476 m

Source: Fieldwork.

The main livelihoods of the people are agriculture and livestock rearing. The main crops grown are maize, beans, potatoes and different types of vegetables (in small quantity). Many villagers work as daily wage labourers in Sikkim. There is a high school in Rammam.

Srikhola Revenue Village

Srikhola Revenue Village (Table 1.4) is situated near the Singalila North Range under the Darjeeling-Sadar sub-division of Darjeeling district, West Bengal. Unlike Gorkhey, Samanden and Rammam, this village is a revenue village, that is villagers have land rights or *patta*. It lies at the confluence of two rivers, namely, Srikhola and Gurdum. The Srikhola River separates it from the Gurdum and Sepi villages. On the eastern side of the village lies the Bijgaon village. On the northern side of the village, there is Srikhola Forest Village. There is a reserve forest on the eastern and western side of the village.

Like Rammam, Gorkhey and Samanden, the main occupations of the villagers are agriculture, livestock rearing and other non-farm employment. Lots of people in the village are involved in tourism activities. The nearest market is Rimbik and Sepi. People have to go to Sepi for availing shared vehicles bound for Rimbik and Darjeeling. The village has one high school.

Rampuria

Rampuria (Table 1.5) is a revenue village situated near the Senchal Wildlife Sanctuary in Darjeeling-Sadar, sub-division of Darjeeling district. The village

Table 1.4 Village profile: Srikhola Revenue Village

Gram Panchayat	Srikhola-Daragaon
Block	Darjeeling Pulbazar
Sub-division	Darjeeling-Sadar
District	Darjeeling
Number of Households	100
Major Religion	Buddhism, Hinduism, Christianity
Altitude	2,200–2,400 m

Source: Fieldwork.

Table 1.5 Village profile: Rampuria Revenue Village

Gram Panchayat	Pubong-Rampuria
Block	Rangli-Rangliot
Sub-division	Darjeeling-Sadar
District	Darjeeling
Forest Range	Senchal
Number of Households	100 (approx.)
Major Religion	Buddhism, Christianity, Hinduism
Altitude	1,600 m

Source: Fieldwork.

is surrounded by the sanctuary, Rampuria Khasmal village, Pubong village and reserve forests under the territorial jurisdiction. The number of households is almost 100. There are small numbers of hamlets in the village named as *Khaling-Gaon*, *Ghale-Gaon*, *Himul-Gaon*, etc. The vast area of the village is under private forest created by the villagers.

The villagers have land rights. However, due to the close proximity to the wildlife sanctuary, an Eco-Development Committee has been set up in the village and the eastern part of the village is under the Forest Protection Committee. The village is 5 kilometres away from the Darjeeling-Kalimpong road. Shared vehicles ply from Rampuria Forest Village and Takdah to Darjeeling every day. People have to come to the Takdah road or the Rampuria Forest Village for availing the facility of shared vehicles. There is one primary school in the village. Most children go to this school or nearby private schools. The village has electricity.

People practise Hinduism, Buddhism and Christianity. The main language for communication is Nepali. The chief occupation of the people is agriculture and livestock rearing. Villagers also get some cash income through MGNREGS either implemented by the Forest Department or the *Panchayat*. People send their agricultural and livestock products through vehicles to the nearby market in Sixth Mile, Ghoom and Darjeeling.

Lalung

Lalung (Table 1.6) is a plantation village under the Mongpu Cinchona plantation area of Darjeeling. The village is surrounded by Cinchona plantations on all sides except the southern side or the Ranju Valley. The Cinchona plantation area on the north and west side is surrounded by the forests of the Senchal Wildlife Sanctuary. The forest in this wildlife sanctuary separates Lalung from Rampuria Revenue Village. The number of households is approximately 55.

The villagers don't have land rights. This village has road connectivity through the Darjeeling-Mongpu road. People go to Darjeeling by shared vehicles. The village has electricity. Like Rampuria the main religions practised here are Hinduism, Buddhism and Christianity. The spoken language is Nepali. The main occupations of the people are agriculture, livestock rearing and non-farm employment.

Table 1.6 Village profile: Lalung Plantation Village

Gram Panchayat	Labdha
Block	Rangli-Rangliot
Sub-division	Darjeeling-Sadar
District	Darjeeling
Forest Range	Senchal
Number of Households	55 (approx.)
Major Religion	Buddhism, Christianity, Hinduism
Altitude	1,600 m

Source: Fieldwork.

Table 1.7 Village profile: Rambi Forest Village

Gram Panchayat	Labdha
Block	Rangli-Rangliot
Sub-division	Darjeeling-Sadar
District	Darjeeling
Forest Range	Gadikhan
Forest Beat	Rambi
Number of Households	36 (approx.)
Major Religion	Buddhism and Hinduism
Altitude	2,200–2,300 m

Source: Fieldwork.

Rambi

Rambi (Table 1.7) is a forest village situated in the Sadar sub-division of Darjeeling district. It is situated in the wildlife jurisdiction of the Senchal Wildlife Sanctuary. Like Rampuria and Lalung, the main livelihoods of the people are farming and other non-farming work to earn daily wages. The neighbouring villages are Third Mile, Lalung and Chatakpur.

The villagers don't have land rights. This village has road connectivity to Third Mile and Chatakpur. People go to Darjeeling by shared vehicles. This village has electricity.

Chapter outlines

Amartya Sen typically invokes diseased or extreme environments as variables affecting well-being opportunities, just as personal diversities (handicaps) affect conversion of resource into valuable functioning. The second chapter, 'Sustainable Development as Freedom', parachutes Amartya Sen into the Eastern Himalayan forests and grapples with the implications of forests becoming an evaluative context for capabilities and capabilities becoming as evaluative space within forests. The chapter turns Sen on his head. It demonstrates how healthy and gentle environments themselves are variation or constraining handicaps. Their protection under forest and wildlife laws constrain services or resources from generating the ability to do valuable things. The chapter compares actual livings and achievements in Singalila's Samanden and Senchal's Lalung. Education and health deprivations are marked in these two villages.

Sen's is not a substantive theory of justice and is a general framework to evaluate unjust social and institutional arrangements. His capabilities approach is under-specified and thus is applicable in a diversity of disciplinary and applied contexts. But Nussbaum's capabilities approach is a substantive justice theory, and she has abstracted ten central human capabilities. These are to be nationally and locally contextualized. The third chapter, 'Justice and the Jungle: Contextualizing the Central Capabilities in

Singalila and Senchal Landscapes', contextualizes three capabilities in Nussbaum's list in Singalila's Srikhola and Senchal's Rampuria. These include 'control over material and political environment' capability, 'practical reasoning' capability and 'affiliation' capability. The chapter assesses thresholds for these three capabilities (Nussbaum, 2011, p. 34).

In Nussbaum's 'Other Species' capability, concern for animals, plants and nature, and living in a relationship with them, is one human capability with instrumental value. This instrumental capability offers some scope for protecting natural entities with which humans have meaningful relationships. But for Breena Holland, the 'natural environment's instrumental value figures' in a 'limited way' in Nussbaum's capabilities approach. Holland seeks to establish these environmental conditions as an independent meta-capability. By meta-capability, Holland refers to the 'sustainable ecological capacity' of a household (Holland, 2008, p. 324). The fourth chapter, 'Sustainable Ecological Capacity of Forest-Fringe Families', empirically operationalizes Holland's concept of 'meta-capability'. The concept of sustainable ecological capacity is measured in terms of forest conditions and community access to forests across Singalila and Senchal's villages. The chapter demonstrates in Senchal the creation of Protected Areas led to the subsequent ban on grazing and restricted fuelwood collection. Further, the relocation to fringe areas devoid of fuelwood stock. This impeded the ability of village households to 'function' because of the compromised meta-capability. However, households in Senchal are able to compensate this disentitlement in terms of their growing connectivity and linkages to markets. This is not the case for Singalila as it still remains geographically isolated.

Nussbaum's and Sen's philosophy of well-being, when verified in an 'isolated' forested Himalayan hamlet, yields interesting well-being and conservation insight. The sixth chapter, 'Our Village Would Have Been Heaven: What the Families of Gorkhey Value', compares the capabilities of families of Gorkhey, a hamlet in Singalila. In the collection, for neighbours and friends, of scarce nutritional resource from the forest, community affiliation is signalled. In Gorkhey's inability to make energy switches, their capability to plan and conceive of what is good, convenient and sustainable is compromised. In the natural-historical observations of Gorkhey folk on the correlation between manure deficiency and bamboo regeneration, their reasoning capabilities assume conservation policy significance. The firm natural historical belief in Gorkhey is that the grazing ban has denied crucial nutrition input into the forest floor in the form of manure. That this has affected bamboo regeneration is further evidence that timber provisioning services are in fact co-produced.

Co-production of services is one way to forge stronger links between ecosystem services and capabilities frameworks. Disruption of human and ecosystem interactions can have well-being consequences. People in Singalila say that wildlife like boars do not find food in such compromised forests and raid village crops. The fifth chapter, 'Our "Other Species" Capabilities and

Capabilities of Other Species', sets 'conservation' context for Nussbaum's other species capability, namely, human-wildlife conflicts. Critiquing human and wildlife conflict studies for their inadequate use of 'attitudes', 'perception' and tolerance, the chapter contends that such 'fleeting' subjective research is blind to adaptive preference (Nussbaum, 2011). The chapter also critiques Nussbaum's argument that wilderness cannot be an excuse for not protecting individual 'sentient' animals that can feel pain and thwarted. In the name of wilderness protection local communities have been disentitled and capabilities thwarted. Differentiating wildlife conservation from urban animal welfare, the chapter compares conflict, capability and adaptive tendency in Lalung and Rammam villages.

Conservation policy and practice must be inclusive of community well-being. But why it should also be inclusive of how well-being needs to be understood. The seventh chapter, 'Conclusion: Inclusion', invokes inclusion in more ways than one. Substantive freedoms to do and be, as a development indicator and outcome, is an infant of an idea in India's forests. From this Eastern Himalayan study, the chapter draws conceptual and policy implications on the capabilities approach to well-being in a conservation and sustainable development context.

References

Agrawal, A. and Redford, K. (2006). *Poverty, Development and Biodiversity Conservation: Shooting in the Dark?* WCS Working Paper. 26.

Bauman, Z. (2000). *Liquid Modernity*. Cambridge Polity Press, p. 36.

Baviskar, A. (1995). *In the Belly of the River-Tribal Conflicts over Development in the Narmada Valley*. New Delhi: Oxford University Press.

Baviskar, A. (1997). Ecology and Development in India: A Field and its Future. *Sociological Bulletin*, 46(2), pp. 193–208.

Bell, M. (1998). *An Invitation to Environmental Sociology*. Thousand Oaks, CA: Pine Forge.

Brockington, D. (2002). *Fortress Conservation. The Preservation of the Mkomazi Game Reserve, Tanzania*. Oxford: James Currey.

Brown, K. and Westaway, E. (2011). Agency, Capacity and Resilience to Environmental Change: Lessons from Human Development. *Annual Review of Environment and Resources*, 36, pp. 321–342.

Chambers, R. (1995). Poverty and Livelihoods: Whose Reality Counts? *Environment and Urbanization*, pp. 173–204.

Deaton, A. (2008). Income, Health, and Well-Being Around the World: Evidence from the Gallup World Poll. *Journal of Economic Perspectives*, 22(2), pp. 53–72.

Ehmke, E. (2012). Ideas in the Indian Welfare Trajectory. Welfare Regimes in the Global South. *Journal Fur Entwicklungspolitik*, XXVIII(1–2012), pp. 80–102.

Fry, P.B. et al. (2017). Monitoring Local Well-Being in Environmental Interventions: A Consideration of Practical Trade-offs. *Oryx*, 51(1), pp. 68–76.

Gadgil, M. and Guha, R. (1992). *This Fissured Land: An Ecological History of India*. New Delhi: Oxford University Press.

Guha, R. (2000). *Environmentalism: A Global History*. New Delhi: Oxford University Press.

Guha, R. and Martinez-Alier, J. (1998). *Varieties of Environmentalism: Essays North and South*. New Delhi: Oxford University Press.

Hands, W. (2009). Economics, Psychology, and the History of Consumer Choice Theory. *Cambridge Journal of Economics*, 34, pp. 633–648.

Harvey, D. (2005). *A Brief History of Neoliberalism*. New York: Oxford University Press.

Holland, B. (2008a). Justice and the Environment in Nussbaum's, "Capabilities Approach": Why Sustainable Ecological Capacity Is a Meta-Capability. *Political Research Quarterly*, 61, pp. 219–332.

Holland, B. (2008b). Ecology and the Limits of Justice: Establishing Capability Ceilings in Nussbaum's Capabilities Approach. *Journal of Human Development*, 9, pp. 401–425.

Keyes, C. and Waterman, M. (2003). Dimensions of Well-Being and Mental Health in Adulthood. In: N. Bornstein, L. Davidson, C. Keyes, and K. Moore, eds., *Well-Being: Positive Development Across the Life Course*. New Jersey: Psychology Press, pp. 477–497.

Khaling, S. and Pradhan, S. (2017). Engaging in Eastern Himalaya-Northeast India: Twenty Years and Beyond. In: A.J. Hiremath, N.D. Rai, and A. Siddhartha, eds., *Transcending Boundaries: Reflecting on Twenty Years of Action and Research at ATREE*. Bangalore: Ashoka Trust for Research in Ecology and the Environment, pp. 44–51.

Krishnan, S. et al. (2017). Beyond Trekker Platitudes: How Forests and Farmers Fare in an Eastern Himalayan Forest Edge. In: A.J. Hiremath, N.D. Rai, and A. Siddhartha, eds., *Transcending Boundaries: Reflecting on Twenty Years of Action and Research at ATREE*. Bangalore: Ashoka Trust for Research in Ecology and the Environment, pp. 36–43.

Lele, S. et al. (2013). Ecosystem Services: Origins, Contributions, Pitfalls, and Alternatives. *Conservation and Society*, 11(4), p. 343.

Loewe, M. (2009, November 20). *Right to Social Security in Development: Facilitating National Conditions: What Action Is Needed?* Intervention at International Symposium "Right to Social Security in Development", Friedrich-Ebert-Stiftung, Berlin.

Mazur, A. (1999). Global Environmental Change in the News Sociological Perspectives on Global Environmental Change: Part Two. *International Sociology*, 13(4), pp. 457–472.

MEA (Millennium Ecosystem Assessment). (2005). *Ecosystem and Human Well-being: Synthesis*. Washington, DC: Island Press.

Mittermeier, R., Robles Gil, P., Hoffmann, M., Pilgrim, J., Brooks, T., Goettsch Mittermeier, C., Lamoreux, J. and Fonseca, G. (2004). *Hotspots Revisited. Earth's Biologically Richest and Most Endangered Terrestrial Eco-regions*. Ciudad de Mexico, Mexico: CEMEX.

Norgaard, R.B. (2010). Ecosystem Services: From Eye-Opening Metaphor to Complexity Blinder. *Ecological Economics*, 69(6), pp. 1219–1227.

Nussbaum, M. (1992). Human Functioning and Social Justice: In Defense of Aristotelian Essentialism. *Political Theory*, 20(2), pp. 202–246.

Nussbaum, M. (2011). *Creating Capabilities: The Human Development Approach*. Cambridge, MA: Belknap Press of Harvard University Press.

Plato.stanford.edu. (2019). *Hedonism (Stanford Encyclopedia of Philosophy)*. Available at: https://plato.stanford.edu/entries/hedonism/#PsyHedf-out Last accessed on 28 February 2020
Rasmussen, D. (1999). Human Flourishing and the Appeal to Human Nature. *Social Philosophy and Policy*, 16(1), pp. 1–43.
Rawls, J. (1971). *Theory of Justice*. Cambridge, MA: Harvard University Press.
Roy, T. (2011). *The Economic History of India, 1857–1947*. New Delhi: Oxford University Press.
Schroter, M. et al. (2014). Ecosystem Services as a Contested Concept: A Synthesis of Critique and Counter-Arguments. *Conservation Letters*, 7(6), pp. 514–523.
Scoones, I. (1998). Sustainable Rural Livelihoods: A Framework for Analysis. IDS Working Paper.
Sen, A. (1984). *Resources, Values and Development*. Oxford: Basil Blackwell, pp. 307–324.
Sen, A. (1987). *The Standard of Living*. Cambridge: Cambridge University Press.
Sen, A. (1999a). *Development as Freedom*. New York: Alfred A. Knopf.
Sen, A. (1999b). The Possibility of Social Choice. *The American Economic Review*, 89(3), pp. 349–378.
Skocpol, T. (1992). *Protecting Soldiers and Mothers: The Political Origins of Social Policy in the United States*. Cambridge, MA: Belknap Press of Harvard University Press.
Steinmetz, G. (2014). The Sociology of Empires, Colonies, and Postcolonialism. *Annual Review of Sociology*, 40(1), pp. 77–103.
Sundar, N. (1997). *Subalterns and Sovereigns: An Anthropological History of Bastar, 1845–1996*. New Delhi: Oxford University Press.
Wimmer, A. (2014). War. *Annual Review of Sociology*, 40(1), pp. 173–197.

2 Sustainable development as freedom

Siddhartha Krishnan and Neha Mohanty

Amartya Sen, in his vast body of work, explored the idea of human freedom. This in the context of development and sustainable development. Solving environmental problems requires the participation of humanity as a whole. Sustainable development as espoused by the Bruntland Commission is about the needs of today's and future generations. A holistic understanding of sustainability demands sustenance of human freedoms that people value. Not just needs-fulfilment. Human beings for Sen are 'reflective creatures' (Sen, 2013, p. 6). They would not want to be denied of their agency to take individual and joint decisions. They would want to partake in decision-making processes about their well-being.

Conservation and livelihood studies, especially of the ecological economic kind, deal with well-being. But focus is on income, valuation and satisfaction. Quantifiable perception surveys are usually employed to generate preferences and satisfaction. Though disentitled and deprived by conservation laws, forest people reveal high utility of the ecosystem and its services. But this signals resignation or habituation to stressful conservation circumstances. Amartya Sen urges that we be suspicious about revealed preferences and perceived satisfaction being objective well-being measures. Hence, utilitarianism for its monistic sense of ethics with utility being its only informational base (Polishchuk and Rauschmayer, 2012; Sen, 1999).

For decades, Sen (1999, 2009) has nudged scholars and policy makers to go beyond income as well-being indicator. Income from ecosystem services is important. But does this income enable capabilities or substantive freedoms to pursue meaningful opportunities? This includes education, health and other valuable local pursuits that Sen refers to in general.

This chapter explores what it means to introduce Sen's capabilities in conservation contexts. The following section strings together an environmental theory of Sen from his thoughts on the subject. A taxonomy of forms or *avatars* the environment assumes in his work. Here the environment serves as contradictory state, public space, constrain space, freedom beneficent, development beneficent, ethical opportunity and sustainable freedom opportunity.

The section that follows this is a critique of ecosystem functions and the Forest Rights Act (FRA) as freedom 'processes' (Sen, 2005, p. 156). For Sen, freedom involves processes, along with opportunities. Ecosystem services or forest rights are public goods and their links with well-being are not linear and vary. As public good or process, services or rights must deal with variations. Sen typically invokes diseased or extreme environments as variables affecting well-being opportunities, just as personal diversities (handicaps) affect conversion of resource into valuable functioning. We turn Sen on his head. Healthy and gentle environments themselves vary in their well-being effects, and can also constrain or handicap. Their protection under forest and wildlife laws constrain services or resources from generating 'the ability to do valuable things' (Sen, 1999, p. 72). The ecosystem services type conservation thinking assumes the services are commonly beneficial. But there is the need for a 'human development' type of reporting 'on the actual lives lived by people, especially by the relatively deprived' (Sen, 1999, p. 73). To 'concentrate on the actual living that people manage to achieve' or 'on the freedom to achieve actual livings that one can have reason to value' (Sen, 1999, p. 73). Actual livings in Senchal's Lalung and Singalila's Samanden forms the next comparative section. We compare nutrition, health and education choices and opportunities in these two forest-edge settlements. These are substantive things, the freedoms to pursue which conservation law thwarts. Our conclusion is post-colonial. We examine how a post-colonial critique of Sen fares in a conservation context.

Sen's environmental theory

In Amartya Sen's development and sustainable development thought, the environment plays seven broad roles, namely, contradictory state, contingent factor, discursive realm, freedom beneficent, ethical opportunity, reasoning opportunity, development opportunity and sustainable freedom opportunity. The environment assumes these seven roles as Sen's thinking crosses over from development to sustainable development. In this journey, freedom goes from needing to be substantive to sustainably so. We use his *Development as Freedom* and *The Idea of Justice* as broad milestones that mark his journey (Sen, 1999, 2009). Following, we discuss each of these avatars the environment assumed in the laureate's work. Capabilities, or substantive and sustainable freedoms that development is to guarantee, plays crucial roles in most of these avatars.

Environment as contradiction

Sen does a Dickens – '*It was the best of times, it was the worst of times*'– in comparing realities of the 20th century with preceding ones. Not just economic change, politics has changed for the good and democracy is 'the preeminent model of political organization' (Sen, 1999, p. xi). Human rights and liberty are key rhetoric. Human life expectancy has increased. Globalization links the world's regions more closely in economic and intellectual terms. Yet, the last century was remarkable for deprivation, destitution and oppression. Environmental and sustainability threats are a contradiction in these good times. Along with poverty, famines, hunger, political freedom and liberty violations, and women agency neglect, prevail. Sen links these contradictions of opulence and environmental degradation. From a consequentialist perspective, private property has propelled economic expansion and prosperity. But unrestricted private property contributes to 'entrenched poverty' and is 'defective in ensuring environmental preservation' (Sen, 1999, p. 61).

Environment as contingency

Without tax regulation, income can accrue as does its environmental consequences. But the environment is also consequential in its 'contingent' capacity. For Sen, incomes and commodities may be 'material basis of well-being' but the uses we make of these is dependent on contingent circumstances. The environment is one source of variation between incomes and the well-being – freedom – we gain. Environmental 'diversities' or variations include climatic conditions like temperature, rainfall and flooding. These variations influence what a person can achieve with a certain income level (Sen, 1999, p. 109). Using forest-based examples, heating and clothing requirements of forest dwellers in Singalila during winter have expenses that a semi-arid forest dweller may not face. A malarial forest 'alters the quality of life' of its inhabitants. So does, as Sen says, 'other environmental handicaps' (Sen, 1999, p. 70). We shall see that even a congenial and resourceful biodiversity hotspot can be an environmental handicap because of policy. Policy that restricts or regulates access, or is indifferent to the spoils of one important constituent of biodiversity – wildlife.

Sen maintains that environmental variations are important for evaluating public action and policy practice around inequality and poverty. Like age and gender, a flood- or drought-prone 'location' can affect conversion of income into capability. Likewise, biodiverse locations are disposed to protection policies that can deprive forest incomes or thwart conversion of income or resources into capabilities.

Environment as discursive realm

The environment is also discursive public realm. Public debates for instance on sustainability and climate change, 'permitted by political freedoms and civil rights', can help formation of environmental values (Sen, 1999, p. 158). Public discussion is a democracy correlate, and cultivating it improves democratic functioning. A well 'informed and less marginalized public discussion of environmental issues' is good for the environment and important for the 'functioning of the democratic system itself' (Sen, 1999, p. 158). Open conservation conversations over *Adivasi* rights and tiger needs then fits Sen's bill of a public discussion. Debates during legislation of the Forest Rights Act on the need to democratize forests were about improving democratic functioning.

Environment as freedom beneficent

The environment benefits from more not less freedom. Critical of Malthusian thinking, Sen nevertheless examines how women emancipation can address the 'world population problem'. Men and women suffer 'general problems of environmental overcrowding'. This links closely with women's 'freedom from the constant bearing and rearing of children' (Sen, 1999, p. 195) in the developing world. Negative links between female literacy and fertility are 'empirically well founded' and observed in many nations. Education empowers women to resist 'continuous child rearing'. Educated women have 'greater freedom to exercise their agency in family decisions' on fertility and childbirth. This is an empirical reinforcement of 'development as freedom' (Sen, 1999, p. 199).

Environment as ethical and critical opportunity

There is a need to look beyond the market for environmental protection. The environment is also an ethical opportunity. Developing environmental ethics can achieve things that are proposed to be achieved through regulation or through taxes and subsidies. Modern 'prodigals and projectors' – Adam Smith's term for wasteful investors – foul air and water. Behavioural restraints are important here. The environmental challenge is also 'public goods' commonly than individually enjoyed. Provisioning these goods requires more than state regulation or social provisioning. Social values must be developed along with 'sense of responsibility' (Sen, 1999, p. 269).

Human behaviour has deteriorated the environment. But contemporary humans do not seek to harm or be 'callous' about interests of future generations. Sen considers the failure to care for the environment sustainably as a 'lack of reasoned engagement and action' (Sen, 2009, p. 48). Negligent and callously obdurate humans can cause environmental catastrophes. Goodwill is not enough, and we need critical scrutiny. Our 'ally' here is reasoning. The environment is a reasoning opportunity. But we need guard against being 'over-convinced' by one's own arguments and be open to counter-arguments (Sen, 2009, p. 48).

Environment as development beneficent

Sen reasons on how to think about environmental problems. His is a 'quality of life' focus that is anti-preservationist. It is 'superficially plausible' that forest cover or species for example are best protected if humans least interfere. But the environment is not intrinsically valuable and gains value from opportunities it offers people. Its value lies in its impacts on human lives. For this unsurprising reason environmental sustainability is defined in terms of enhancing quality of human life. We are 'indebted' to Gro Brundtland's sustainable development formula, Sen says. Her 1987 report defined 'sustainable development' as 'development that meets the needs of the present without compromising the ability of future generations to meet their own needs'. Sen goes on to critique what is to be sustained but at the same time appreciates the understanding the commission generated that 'the value of the environment cannot be divorced from the lives of living creatures' and that 'the environment is not only a matter of passive preservation, but also one of active pursuit' (Sen, 2009, p. 249).

While development can be destructive of the environment, it can also be constructive too. Development can enhance effective intervention and reasoning. Spreading quality education increases environmental consciousness. Active and informed media can increase environmental awareness. Considering development as human freedom increasing 'brings the constructive agency of people engaged in environment-friendly activities directly within the domain of developmental achievements'. Development is fundamentally an empowering process, and this power can protect the environment and 'not only to decimate it' (Sen, 2009, p. 249).

Environment as sustainable freedom opportunity

Is the conception of humans as needy contemporary and future entities 'an adequately capacious view of humanity?' Sen asks (Sen, 2009, p. 250). People have values and not just needs. They 'cherish their ability to reason, appraise, choose, participate and act' (Sen, 2009, p. 250). A needy perspective of humanity is a 'meagre' one. Even Robert Solow's extension of sustainability as the next generation be left with 'whatever it takes to achieve a standard of living at least as good as our own and to look after their next generation similarly' is not a broad view of humanity (Sen, 2009, p. 250). Sustaining living standards is not same as sustaining freedom and capabilities to pursue valuable opportunities. But reasons for valuing some opportunities need not always contribute to our living standards and interests. Sen illustrates this 'altruism' through species conservation. A species facing extinction may not enhance our living standards but its conservation is important. Paraphrasing Sen, there would be no contradiction if the *Sherpa* said: 'My living standards would be largely, indeed completely, unaffected by the presence or absence of red pandas – I have in fact never even seen one – but I do strongly believe that we should not let those pandas become extinct, for reasons that have nothing much to do with human living standards'.

In needing to go beyond these individual motivations and understand social commitments to populations vulnerable to environmental adversity like climate change, Sen approximates environmental justice reasoning.

Flooding in Bangladesh from sea level rise and its threats to precarious populations can influence the thoughts and actions of people who do not face these threats. Here the environment is a reasoning opportunity that can get one to reflect upon roles of temperate northern nations in carbon emissions that raise sea levels in third-world coasts. But Sen does not elaborate upon this glaring 'un-freedom'.

Back to the species question, Sen does refer to the *Sutta-Nipata*. In it, the Buddha argues our animal responsibility arises 'precisely because of the asymmetry between us, not because of any symmetry that takes us to the need for cooperation'. Since 'we are enormously more powerful than other species, we have some responsibility towards other species that connects exactly with this asymmetry of power' (Sen, 2009, p. 205). Our conservation efforts need not always be connected to on our need fulfilment. Some of them 'turn precisely on our sense of values and on our acknowledgement of our fiduciary responsibility' (Sen, 2009, p. 251).

If our worth is more than being needy entities, but also lies in our freedoms, then sustainable development has to be reformulated. Our need fulfilment must be sustained, but also our freedoms, including freedoms to fulfil our needs. Sustainable freedom is a reformulation of Bruntland's sustainable development. It is the 'preservation, and when possible expansion, of the substantive freedoms and capabilities of people today "without compromising the capability of future generations" to have similar – or more – freedom'. Sen invokes the 'medieval distinction' between patients and agents. We paraphrase again from the shoes of a Dukpa. We are more than 'patients' whose needs be treated, but also 'agents' 'whose freedom to decide what to value and how to pursue what we value can extend far beyond' our own farming and forest foraging interests and needs. 'The significance of our lives cannot be put into' any protected area context, or our nutrition and energy need fulfilment. Our manifest needs as patients are important, but cannot conceal the importance and relevance of our 'reasoned values' as agents (Sen, 2009, pp. 251–252).

Ecosystem services and forest rights as public goods

Recalling Sen's capability framework, it is an alternative development approach to income and satisfaction as well-being indicators. The capability framework makes two main claims. First, one must have the freedom to achieve well-being. Second, that we understand this freedom as people's capabilities or their real prospects to do and be what they value. Freedom as capability refers to the extent to which a person is free to choose certain levels of functionings. These are 'beings' and 'doings', such as being well-nourished or educated. And functioning differs between people in a hamlet. And between people in different hamlets. There is no necessary commensurability between ecosystem and human functioning. Going by de Groot et al.'s (2002, p. 3) utilitarian definition, 'ecosystem function is the capacity of natural processes to provide goods and services that satisfy human needs directly or indirectly'. But robust service delivery need not translate into easy serviceability. A provisioning service like fuelwood, or 'cultural service' like recreation can only correspond to – or 'service' – people's well-being levels. These depend on their social, cultural, economic and geographical locations. A diversity in circumstances prevails in a community aggregated as homogenous.

Robust ecosystem services, however measured, do not automatically ensure well-being for forest-edge farmers. Any calculated or derived income is no guarantor of well-being, nor is it a harbinger of 'development'. We are dealing with development as freedom here. Well-being as human flourishing where farming in forests is itself valuable. Not development as singular 'needy' end ensured by farming income. Not revealed satisfaction from cultivation money earned, or roof rain-proofed. We cannot reduce well-being to basic necessities. We cannot allow conservation discourse to reduce development to such basics. That is an unjust and dignity denying discourse to retain power over 'pristine' ecosystems. It assumes aspiring and flourishing people are a threat to forests.

Using another instance, pollination is a support or biodiversity service for pea cultivation. But this crucial micro-activity does not exhaust well-being. Macro-activity of what the farmer is able to valuably do and be with pea income is well-being. Including being aware of the micro-activity of pollination. In rhetoric, an ecosystem provisions fuel and fodder, regulates soil and biologically supports farming. And protect it, we must. But why would you protect the ecosystem from access by its immediate beneficiaries, the forest-edge farmers? Or why would you allow farmers to benefit from pollination and soil regulation but not allow them to sell serviced crops?

Forest rights, service access and control are important to convert services into capabilities. The same is the case with development liberties – roads, hospitals, market access and schools. But these liberties or rights themselves are valuable only to the extent they enable functioning, choice and flourishing. These opportunities and abilities that differ within and between communities. You can linguistically convert natural resources into ecosystem services. And economically convert services into dollar value. But if services cannot be converted into valuable functioning, they remain utilitarian rhetoric. We can assume ecosystem services, like natural resources, to be a Rawlsian public good. But in the absence of other public goods like rights and liberties, its public value is in doubt.

The Forest Rights Act, 2006 (FRA), entitles people with individual and collective land and conservation rights. Through entitlement, it can enable meaningful freedoms or capabilities to choose what to grow, sell and earn and what valuable things they can convert into income. But the West Bengal state did not implement the FRA's provisions in Darjeeling as late as September 2018 because *Panchayat* elections have not been held in the Gorkhaland Territorial Administration (GTA) area. *Gram sabhas* are important rights verifying institutions under the FRA. But even if implemented, the Act by mandating conservation responsibilities raises freedom doubts. It entitles communities with 'sustainable' development rights, but does not treat 'sustainable' development as freedom itself. The FRA treats *Adivasis* and other dwellers as passive need recipients and not active agents. Remedying historical injustice is one great thing. But infantilizing people is another insufficiently democratic thing. You need a valuable crop to be pollinated, and have the freedom to choose whether you will subsist off the yield or sell it. And invest the income for your daughter's desired education as a software engineer. And not by legislated necessity, a territory tethered sustainable forest produce harvester.

Sen's freedom perspective involves process, namely, action and decision freedom. Unfreedoms can arise from inadequate process, in our case from lack of political and civil rights in forest. Ecosystem processes – water and nutrient cycling, energy flow – foster ecosystem functioning, provide services. The FRA as democratic process provides rights. These then are political and ecological processes that provision rights and resources. But what opportunities do these rights and resources enable? Sen's freedom perspective also involves opportunities or capabilities. And unfreedoms also arise from lack of opportunities in forests to live secure, healthy and nourished (Sen, 2009).

Now rights and services are public goods. But the 'broadening of the informational focus from incomes to primary goods' is inadequate in dealing with 'variations'. The relationship between 'income and resources, on the one hand, and well-being and freedom, on the other', varies. Among variations Sen identifies like personal heterogeneities and social climate are environmental diversities (Sen, 1999, p. 72).

Lalung and Samanden: an *actual living* report

Lalung is an accessible Cinchona plantation settlement bordering the Senchal Wildlife Sanctuary. Samanden is a forest settlement located more remotely near the Singalila National Park (see Figure 2.2). We interviewed 44 households in Lalung and 27 in Samanden (see Figure 2.1). In all, 11 interviewees were women.[1] What actual dietary, health and education capabilities do farming families manage to achieve? What freedoms do they have to live valuably nourished, healthy and educated? Let us first compare dietary and nourishment achievements in Lalung and Samanden.

The primary source of carbohydrate is rice and *roti* for Lalung families, a staple consumed at least once a day. Vegetables including spinach, peas, cabbage, cauliflower and beans meet their regular nutritional needs. Beet, brinjal and lady's finger are occasional choices. Potatoes form their main starch source. Protein source is chiefly *dal* (lentil), but once a week they also consume chicken, beef, pork and fish. Some cook and consume meat twice or thrice a week. Being nearer to urban Darjeeling, meat is more accessible.

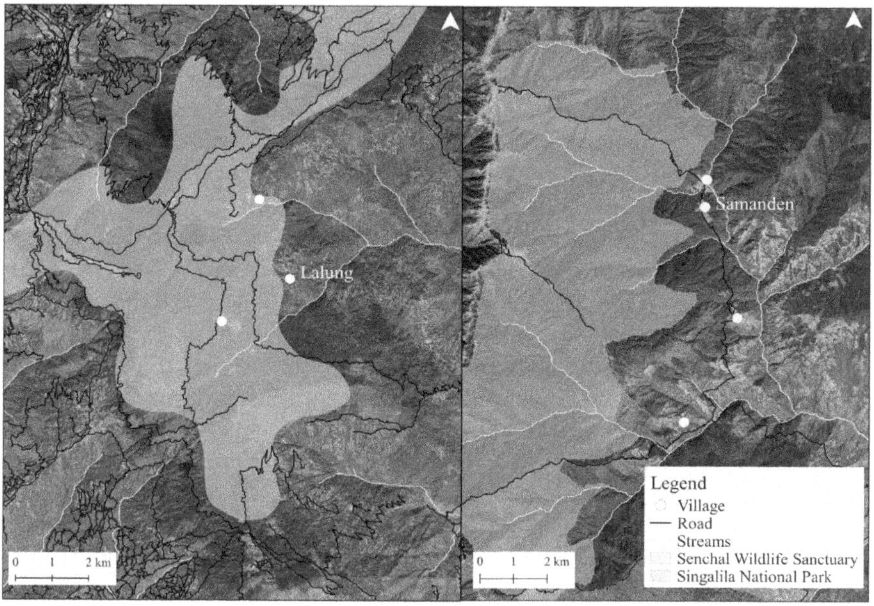

Figure 2.1 Lalung and Samanden: the topography

Source: Generated by Abhishek Samrat.

Figure 2.2 Samanden: the landscape

Source: Roshan Rai.

Lalung responses suggest a more positive dietary discrimination for pregnant women and children as compared to old people. People also report stark changes in their ancestral diets. Up until two decades ago, maize was a primary carbohydrate source. It was widely grown and consumed. The cultivation and consumption of maize has drastically reduced. The sixth chapter details how wildlife raids have forced people to give up their choice of growing and consuming maize and peas.

The dependence on market for food has increased. Constant yield depredation has meant dependence on market for dietary needs. Besides, *dal* is always bought from the market. Potato cultivation persists. In the last three to four years, new crops like cabbage, coriander, carrot, cauliflower, chili, radish and beet are grown on a small scale. Wildlife are not the only 'environmental' factor influencing sowing decisions and harvest achievements. Seasons are a major cause too. People's dependence on the market increases in winter. Crops yield lessens in comparison. Meat consumption increases in winter as does consumption of soup and the occasional porridge.

Families also consume wild vegetables like *ningro*, *sisnu* and mushrooms that they access from plantations in nearby Mungpo. Their availability peaks during April and July. These medicinal plants are not as abundant as before. People recall reduction over a decade. Agricultural decline parallels livestock decline. The latter due to grazing access restrictions. Nutritional options and dietary patterns have changed. A range of reasons is afforded for this change. Manure shortage and increased use of chemicals has reduced soil fertility. And there is extreme weather conditions and wild animal menace. But wild animal raids and wildlife laws that render people helpless in responding to raids in ways they value is as significant a cause as dietary change. Even as elders struggled with cropping choices, the younger generation appears to have made its dietary choices. The latter do not prefer eating maize. They are not interested in maize cultivation given the hardship and risks. Also, since there is more effort in growing crops whose yield is declining, it is demotivating.

Women claim that changes in nutritional patterns have rendered them weaker than men. Forest laws have reduced medicinal choices. People collected medicinal plants like Harchur (*Viscum articulatum*), Chirauto (*Swertia chirayita*), Kali Jhar (*Eupatorium adenophorum*), Pakhanbet (*Berginia ciliata*), Gheu Kumari (*Aloe vera*), Betlouri (*Costus speciosus*), Buro okhati (*Astilbe rivularis*), Banmara (*Eupatorium adenophorum*) and Titeypati (*Artemesia vulgaris*). Medicinal plants were sourced from both the Mungpo plantation and the Senchal sanctuary. Deforestation and increased competition combine in reducing medicinal plant availability. Further, people's medicinal collection choice is vulnerable to punitive thwarting. They face fines up to 1,000–1,500 INR when caught. This has resulted in increased dependence on allopathy. Also a reduced freedom to be sustainable, as medicinal plant collectors or users. Owing to lesser access to and availability of medicinal plants and increased competition, people focus on the needs of the day. Tomorrow's needs are ignored, and people collect whatever they can find before someone else does.

There is not a stark difference in dietary practices of Samanden. Dietary staple is rice, *dal* and seasonal vegetables. *Rotis*, meat and eggs are occasionally eaten. Wheat and meat need to be procured from town markets that are far away. Like Lalung, potato is Samanden's favourite starch. Families include green leafy vegetables like spinach, *gundruk* and *sinki* when they are available on season. Upon occasion, squash, beans, cauliflower and beets are cooked. However, unlike Lalung, more respondents from this region grow and consume maize. But the soil was not yielding as it did, and boars and bears were unyielding in their maize raids.

Dal and eggs are a popular choice of protein. Meat is cooked once a month on average. Given market distance and affordability, Samanden, unlike Lalung, cannot afford regular meat consumption. People also rely on the market for rice, oil, spices, salt, and upon occasion, even vegetables. Dependence on the market has increased in the past decade.

Except for them consuming soups in winter, respondents claim little or no seasonal variation in diet. Samanden households practise positive diet discrimination for pregnant women and nursing mothers in terms of both food quantity and nutritious diet.

Farmers report decreasing consumption of wild mushrooms, bamboo shoots, nettles and fiddlehead leaves. They recall 'the olden days' when they could bring these home in large enough quantities to meet their nutritional needs. Forest regulations are strict, and the department restricts collection. Opportunities to collect valuable nutritional sources were receding. Many respondents even state that they are 'scared to enter the forest'. They even refrain from entering accessible forest patches as 'nothing is available'. A few people collect medicinal plants like *harchur, chirauto* and *budo okhati*. But use has drastically reduced and people also value allopathy and prefer going to clinics and dispensaries located at distances. This is an interesting contrast to more urban Lalung which values medicinal plants. The reduction in access to forests in both villages compromises people's choice to procure medicinal plants they used for healthcare.

Taking stock of dietary capabilities, Lalung and Samanden do not experience hunger or malnutrition. Their nutritional requirements are largely met. But they had to switch from nutritive maize to a simple carbohydrate like rice. Chemical fertilizing has reduced soil capabilities. Wild animals depredate crops. Soil and animal are unyielding. Precarious yield increases market reliance for rice, potato and vegetables. Lalung relies more on the market for vegetables. Samanden tones down any meat need as the market is far flung.

Now markets are bad words for conservationists and their political-ecological critics. There are many reasons why. But a common worry is 'modern' contamination of traditional sustainable living. Calculation replaces communitarianism; people will poach tigers and over-pluck fruits. The market's contribution to economic growth aside, the freedom to interchange goods is significant for Sen. The 'freedom to exchange words, or goods, or gifts does not need defensive justification', 'they are part of the way human beings in society live and interact with each other' (Sen, 1999, p. 6). But sole reliance on markets would be an imposition on people. People's food habits regularly and decisions about what they buy or grow cannot be imposed. And conservation policy cannot oppose market access. People and their expansion of freedoms, including freedom to grow or buy, need drive this change and not markets or conservation laws. Lalung's market conversations are more intimate. Market provisions food grown earlier.

What people value

Lalung and Samanden value education and health. People value the former's role in achieving better outcomes for the latter. This value does not arise in achievement but from absence. There is no hospital or dispensary in Lalung. The nearest health facility is in Jor Bungalow, which is at a distance of 12 kilometres from the village. The other option is Darjeeling which is 22 kilometres away from the village. There is no hospital or dispensary in Samanden either. People need to go to neighbouring Sikkim to access health facilities.

In Lalung, people consider lack of education or lack of concern by the educated as responsible for the non-establishment of health facilities. 'Public are not demanding it', 'lack of interest', 'if villager will be united then possibility' are people's responses to why the administration has not set up a dispensary or hospital. A middle-aged *Rai* farmer links health and education deprivations:

> There is a primary school but no high school in the village. And, these two are the main problems of our village. Our village is a small one; most of the people who are educated do not stay here. Therefore, the voice that we raise for hospital and school does not reach the govt. department. This is a long procedure. Our village is small. We have not been able to reach to that level so far.

If hospital access is development aspiration, then conservation policy depriving access needs a strong democratic dose.

Lalung's not reaching 'that level so far' is a figurative development lament. But having to reach far is Samanden's literal lament. A *Sherpa* farmer provides priority sarcasm, 'we don't have proper roads let alone a hospital'. Sounding hopeless, he says, 'I don't think there is ever going to be a hospital here'. A *Tamang* woman responds to a query about hospitals, 'At the Sikkim side. People carry the sick person to that place'. For a *Rai* farmer, apathy and literacy combine when he says, 'there was no one interested and no one was educated to go to ask for it'. How much people value hospitals is evident to a question of whether people use medicinal plants: 'No I prefer hospital', says a *Chettri* elder. A *Rai* says, 'people just come for survey but don't take any further initiative'.

Like Sen links education and women's freedom to exercise agency in the family, Lalung and Samanden folk believe education empowers community agency. Being educated builds health awareness and the urge to make demands of authorities. Lalung's residents desire a high school for their children. A *Rai* resident is worried: 'We have these kinds of problems here, there is no school, a proper one. They have to go so far after class five and children are small'. The major problem is children 'have to walk almost 7 kms to go to school', a resident says. A *Rai* farmer makes an expansive and logical point: 'Not only will our village children but it will be easy for other children from our neighbour villages like Lalu, Ranju valley, Chayang dara, and Rambi'. 'There are only two primary schools in the two villages. It would be better if there were one high school in one village and one primary school in another village'.

If the presence of two secondary schools in their village worries Lalung, the absence of two things worry Samanden. Residents miss a dispensary and a primary school. Others list roads and electricity. A *Rai* lady get pragmatic and pessimistic: 'Even if there will be a secondary school I don't believe there are enough students'. 'I want to send my child to a good school far from here', she says.

Health capability deprivations are severe in Lalung and Samanden. People value education. They aspire for their children. They desire that children not take up agriculture but opt for education instead which guarantees lucrative jobs. In Samanden, the precariousness of farming lives is prime reason for valuing education. Lalung and Samanden value tourism as definite economic freedom. Trekking-related home-stay tourism in Samanden has been active for a long time as compared to Lalung where it is being slowly introduced. In Lalung people are keen about such eco-tourism because of the economic benefits. Also as it offers scope to learn new things from tourists and exchange knowledge. In Samanden, people value tourism. It is beneficial as their vegetables find a market, they find jobs and income becomes available for children's education. The sixth chapter on neighbouring Gorkhey discusses tourism's value as economic opportunity.

As a post-colonial phenomenon, sustainable development remains a form of rule. A new way in which rule is legitimized and inequalities reproduced. Though there seem to be deeper meanings to the term, the end goal of sustainable development remains economic growth (McMichael, 1996). Sen's capability approach is a departure from this understanding. It focuses on human lives and not simply the resources that people have in the forms of ownership and use. For Sen, freedom is both the 'primary end' and the 'principal means of development' (Sen, 2010, p. 10). The presence of freedoms has a positive causal relation to sustainable development. Its absence has positive causal relation with deprivation. This is a fundamental shift in emphasis from means of living to the expansion of freedom. The capability framework asks how people envision sustainable development. People are not treated as mere spectators and recipients of change. So how do Lalung and Samanden envision sustainable development? Evaluating the situation objectively, it is unfair to place the onus of sustainability on forest dwellers and dependents. They already live in a resourceful yet stressful environmental context. They were not active participants in bringing about this 'negative' policy circumstance.

Despite their disentitlement and development deprivations, people reason about forests, from enhancing forest cover to conserving forests for animals and humans. Chapter five affords deeper insights into Lalung's sentiments about wildlife. Like Lalung's residents, Samanden residents too state wildlife protection is important and that forests are important for wildlife. A common response from residents from the two settlements to the question of development problems is 'everything is fine'. Sen often warns us about how people can, through years of harshness of circumstances, not desire what they can ever expect to achieve. These are adaptive preferences of people in Samanden and Lalung. They arise when freedom of residence is discussed given that they have no land rights. There are no schools, roads and electricity. But other than that everything is fine! Adaptive preferences to conservation deprivations is dealt with at length in the seventh chapter.

In his capabilities approach, Sen often dwells on the importance of women and agency in development precisely because their agency has a direct impact on their well-being and also of their children. Sen writes that the agency of women is dependent on a number of aspects, namely, education, their earning capacity, economic prowess outside the family, property ownership among others (Sen, 1999).

Almost the same number of women were interviewed in Lalung as in Samanden. So, to what extent do women in both regions have and exercise their agency in their well-being? One clear distinction in both regions is how women partake in organizational affairs, both government and otherwise. Women in Lalung are active members of self-help groups and also voice their suggestions in various organizational meetings. Women in Samanden, however, do not actively participate in organizational meetings. They attend these meetings only when their husbands are not attending them and additionally they do not speak at these meetings. 'We only listen, they talk'.

So, women in Lalung have more agency than women in Samanden? In the aforementioned context, perhaps yes. But as part of a social group, women in both regions are structurally disadvantaged. More so as women who are forest dwellers and dependents. When asked whether women play any important role in the village community, several women from Lalung answer in the affirmative, saying that they help in marriages and other such events. Responses from men to the same question were a bit different. They say that the most important role that they play in their community is building roads and involving themselves in organizational affairs.

Moreover, of problems that the respondents state that they commonly face, like, water, paucity of medicinal plants and even fuelwood, women are the more vulnerable lot. A middle-aged woman from Lalung articulates this. 'We have to think about how to fetch the water and all that', she says, 'how to bring the vegetables from the jungle'. Another young woman from Lalung echoes this sentiment, 'the women are more affected than the men. Yes, because they are the ones who have to work in the home as well as outside. . . . Because mainly the women have to do most of the work'. Another middle-aged woman from Lalung voices the problems that they face owing to changing landscapes on account of being women. 'Now the farming is also not good and the vegetables also do not grow well in the jungle so because of that you do not get proper water supply. So the women have to suffer more'. Hence, when it comes to division of labour in a family and as a community, not only is the nature of it more gendered in Lalung, but also women end up with a greater share of the labour.

Women in both Lalung and Samanden report facing a lot of difficulty when collecting Non-Timber Forest Produce (NTFP). They also state that problems related to health and roads impact them more owing to their prefixed roles as women. They are more prone to sickness, they say. One positive practice is the diet discrimination for pregnant women and nursing mothers in both regions where their nutrition is prioritized.

It is interesting to see what land ownership means to certain women in both Lalung and Samanden who have come to live here post-marriage. Most of them say that despite spending lesser time in the region as compared to their husbands they would feel threatened in case they were asked to relocate. However, their belongingness to this region is often spoken of in relation to their husbands or in some cases their sons. 'He [son] does not like staying in this place, since there is no man here he is staying here', says an old woman from Lalung. Another middle-aged woman from Samanden who arrived there after her marriage says her family can always relocate to her native Daragon in Sikkim in case they are asked to move. Furthermore, most of them reported that they do not want their children to stay at home but pursue education outside. 'They cannot do farming and survive so we want to manage it somehow and send them to the city. We have that in our heart', says an old woman from Lalung. The men however wished for their children to come back home. Home for both men and women seems to have a different connotation.

Women in Lalung and Samanden experience agency and capability deprivations in a stressful conservation context as compared to men. However, their preference that children study and leave the forest suggests that they aspire for their children's agency – an agency that education and income can usher in.

Conclusion: community and deprivation

Unlike Martha Nussbaum, Sen refuses to endorse a list based capabilities approach for it becomes dogmatic. He argues that people use capabilities for different reasons and purposes. Whatever our focus is, be it evaluating poverty or specifying certain basic human rights is not 'independent of what we are doing and why'. Further, social conditions and their priorities vary. One cannot afford to follow a predetermined 'fixed forever' list of capabilities as there is a need for public reasoning and deriving context-appropriate lists (Sen, 2005, p. 160).

In stressful conservation contexts like Lalung and Samanden, where farmers precariously sow and harvest, or do neither, is Sen's non-listed and under-theorized capability approach adequate? Sen's approach is pro-poor and focuses on the agency of the marginalized. But it remains embedded within a post-colonial neo-liberal set-up, intrinsically hegemonic and oppressive (Comiling, 2014). Sen's capability approach vouches to be participatory and views poverty as capability deprivation. But it is not decolonized. The poor or the marginalized thus remain captive to the resilient capitalist system. Sen provides no escape. The freedom that Sen wishes that the marginalized achieve thus remains inherently limited. Sen does little to critique the neo-liberal capitalist system that causes global injustice. Further, his conception of capabilities is heavily focused on individual agency and privatization (Comiling, 2014). In a conservation context like the Himalayas, Sen's emphasis on individual agency is problematic. In the famed extremities of the Himalayas, well-being depends on cooperative labour and moral economic obligations. The paucity of a communitarian angle in the capability approach framework makes it inadequate.

Sen reiterates public reasoning and participation by the marginalized to expand their freedoms. 'But it is unrealistic to expect them to be capable of participating when their collective mode of living, means of support, and common experience of deprivation are not duly recognized and their self-organization is not encouraged' (Comiling, 2014, p. 13).

There is also ambiguity in Sen's insistence about people's capabilities they value. This as opposed to Nussbaum's listing of central capabilities. It is fraught with problems in a stratified country like India. What if some people value killing or harming other people (Gasper, 2002)? How then does one reconceptualize capabilities in a post-colonial perspective and a conservation context?

A 'communitarian agency' example lies in Chiapas, Mexico. The Zapatista uprising of 1994 took the world by storm when the indigenes of Chiapas declared war on the Mexican state. They organized under the Zapatista Army of National Liberation (EZLN). Demands included 'work, land, housing, food, health, education, independence, liberty, democracy, justice and peace' (Speed, 2014, p. 82). The Zapatista movement was instrumental in gaining constitutional protection for the Mayan indigenous people. This was an instance of people exercising their agency to the fullest and actively taking well-being decisions as a community.

The Zapatistas practice horizontal autonomy. They took charge of health, education, sustainable agricultural and ecological systems. They also advocate equitable gender relations. The communities of Chiapas grow the traditional Mayan corn in a plot or *milpais* regarded as a place of heritage and learning. This generation's knowledge of sustainable agriculture is preserved, protected and reproduced. The 'Mayan schools of dignity' are autonomous schools that teach their children traditional knowledge, culture and heritage. Over the last two decades, the Zapatistas have fought structural violence and socio-economic violence wrought by colonization and globalization. They protect their relationship with the forests and their land.

When it comes to locating capabilities approach in the indigenous well-being context, a research study in Australia provides some answers (Sangha et al., 2015). As per the study, the well-being of indigenous people in Australia and around the world can be linked to the usage and valuing of natural resources. It proposes an integrated well-being model by incorporation of values such as cultural identity which in turn are linked to people's capabilities involving natural systems (Sangha et al., 2015). While it is a welcome move to apply the capabilities approach in forest contexts, whether it is capable of being uniformly applied to all forest contexts and whether it is holistic remains ambiguous.

Post-colonialists argue that Western theories of capitalism and modernity are relevant, but inadequate in understanding the post-colonial world. Despite Marxist influence, they critique the universalist assumptions of historical materialism. They question its assumption that capitalism would expand and function uniformly. Taking the west as the norm for the rest, and projecting enlightenment and Marxist categories 'disregards and silences the realities and experiences in the postcolonial world'. But in highlighting the post-colonial world's distinctiveness, post-colonial theorists deny the 'universal validity of emancipatory norms such as justice, democracy and human rights'. These norms are of common universal interest and 'shared by all human beings irrespective of culture, race, gender, sexuality, religion, or other differences' (Dhawan, 2018, p. 10).

Post-colonialists have understandably not spared Sen and Nussbaum either. His is an under-specified universalism. Hers is an over-specified universalism. We discuss these critiques in the third and sixth chapter. Suffice to say that as universalizing theory, capabilities fares better than utilitarianism, that other famous universal. Its pleasure and pain type satisfaction measurement of forest dweller well-being has been insensitive to powerless habituation to conservation deprivations. The principle 'greatest amount of good for the greatest number' may form consequentialist ethics. But its discriminatory consequences have been pain in the forest's edges and pleasure urban centres. Large mammals and biodiversity hotspots have been protected to the applause of the nation's middle and rich classes. But in this aggregating moment, forest minorities have been deprived of public goods.

The rationale for encouraging a universalizing discourse of capabilities is to counter another discourse that has universalized local forested spaces in the south – forest and wildlife protection. For, the Indian Forest Act is colonial as we saw in the introduction chapter. The Wildlife Act has a definite

American and European stamp. It was enacted after international conservation organizations and wilderness aficionados lobbied. People who live near national parks and wildlife sanctuaries are a protected area precariat. Policies with European and American antecedents deprive the forest dwellers of the substantive freedom they need to pursue valuable lives individually or commonly.

Note

1 All interviews mentioned in the book were conducted in 2016.

References

Comiling, K. (2014). A Postcolonial Critique of Amartya Sen's Capability Framework. *Perspectives in the Arts and Humanities Asia*, pp. 4–15.
de Groot, R., Wilson, M. and Boumans, R. (2002). A Typology for the Classification, Description and Valuation of Ecosystem Functions, Goods and Services. *Ecological Economics*, 41(3), pp. 393–408.
Dhawan, N. (2018). *Marxist Critique of Post-colonialism.* Marx from the Margins: A Collective Project, from A to Z.
Gasper, D. (2002). Is Sen's Capability Approach an Adequate Basis for Considering Human Development? *Institute of Social Studies*, p. 28.
McMichael, P. (1996). The Development Project. In: P. McMichael, ed., *Development and Social Change*. Los Angeles: Cornell University Press, pp. 63–68.
Polishchuk, Y. and Rauschmayer, F. (2012). Beyond "Benefits"? Looking at Ecosystem Services Through the Capability Approach. *Ecological Economics*, 81, pp. 103–111.
Sangha, K.K. et al. (2015). Application of Capability Approach to Assess the Role of Ecosystem Services in the Well-Being of Indigenous Australians. *Global Ecology and Conservation*, pp. 446–448.
Sen, A. (1999). *Development as Freedom.* New York: Alfred A. Knopf, pp. xi–269.
Sen, A. (2005). Human Rights and Capabilities. *Journal of Human Development*, pp. 156–160.
Sen, A. (2009). *The Idea of Justice*. London: Penguin, pp. 45–252.
Sen, A. (2013). The Ends and Means of Sustainability. *Journal of Human Development and Capabilities*, 14(1), pp. 6–20.
Speed, S. (2014). Zapatista Autonomy, Local Governance, and an Organic Theory of Rights. In: *The World of Indigenous North America*. Abingdon: Routledge.

3 Justice and the jungle

Contextualizing the central capabilities in Singalila and Senchal landscapes

Sourya Das, Annesha Chowdhury and Neha Mohanty

Introduction

The term 'justice' has different connotations in various contexts, including that of conservation. It has been theorized in a plethora of ways by many different philosophers. For the wildlife conservationist and natural scientist, protection of flora and fauna takes precedence in the justice discourse. For the social scientist, the well-being of forest dwellers and dependents, their relationship with forests and societal structures are significant. Owing to the complex issues that natural areas in India pose, adopting a holistic theory of justice is of paramount importance. This for both a comprehensive analysis of issues, and seeking context-specific policy solutions.

In the context of conservation, law becomes an important instrument of justice delivery. Hence, it is also important to explore the conundrum of law and violence in this context. Law as we understand it is designed to mitigate or curtail a Hobbesian state of nature. If we look at the social contract imagination, creation and preservation of society as a collective is bound by laws (for the society's longevity and well-being), which would not otherwise exist. Yet, law too is violent. So how does one arrive at the limits of law's violence and judge when this is illegitimate? Investigating this paradox in the heart of conservation law and policy is indispensable in arriving at a holistic theory of justice. This is relevant more so in the context of forest-dwelling communities in India and other dependents for whom conservation laws have caused violence in the manner of eviction and denial of access to forest resources.

In the varied discourses of justice, its accessibility has remained a pertinent theme. Franz Kafka's parable *Before the Law* is an apposite lens to view the accessibility of justice to civilians in India. A man journeys from the country to gain access to law but is stopped by the doorkeeper of law for years. When he turns old and is on the verge of death, he is informed that he was the only one who could gain access to that door which now shall be shut. Such is the state of the judicial system in India. Even though one has access to courts, one is unable to become the recipient of justice. *Adivasis* too have access to courts. However, this isn't sufficient for them to be recipients of justice. They have entitlements under Forest Rights Act (FRA), 2006, but when forests are de-notified, their rights are curtailed. Their access to justice is thus hindered.

Amartya Sen's capabilities approach, although useful, is not a substantive theory of justice. It is a general framework to evaluate unjust social and institutional arrangements. In case of Sen, being abstract helps but it also limits. His capabilities approach is vague and under-specified and thus applicable in various disciplinary and applied contexts (Holland, 2014). This, however, is not an apt tool for analyzing complex conservation questions in India especially when it comes to treating the conundrum that the post-colonial neo-liberal world presents. The conundrum is this – How do you expand people's capabilities in a neo-liberal capitalist system that is intrinsically hegemonic and oppressive and causes global injustice (Comiling, 2014)?

Nussbaum's capabilities approach, on the other hand, is a far more substantive justice theory. Unlike Sen, who prescribes against listing of capabilities, Nussbaum has abstracted ten central capabilities that can be nationally and locally contextualized. Her famed list of ten central capabilities was mentioned in the introduction. To recall, these are life; bodily health; bodily integrity; senses, imagination and thought; emotions; practical reason; affiliation; other species; play; and control over one's environment (Nussbaum, 2011).

Nussbaum, through her approach, does a comparative assessment of life quality. In addition to this, she also theorizes social justice through the key question of 'what each person is able to do and to be'. In this case, it is not about the cumulative well-being of a group but of each individual as an end. The focus is on choice and freedom. The 'crucial good societies should be promoting for their people is a set of opportunities, or substantial freedoms'. The choice of exercising them or not thus remains with the people (Nussbaum, 2011, p. 18).

If law is an important instrument in justice delivery, the root of law in India is the Constitution. The Fundamental Rights as enshrined in the Constitution are basic universal rights based on dignity. The basis of Nussbaum's capabilities approach is also dignity and social justice (Nussbaum, 2000). Justice and dignity have been at the root of environmental struggles in India, especially access and control rights over forests. This chapter focuses on three capabilities from Nussbaum's list, namely practical reasoning, affiliation and control over one's environment. The first two are architectonic capabilities. These capabilities are central to being human and inform and organize all of the other capabilities (Nussbaum, 2011). The rationale behind using the third one is its relevance as an architecton in depriving conservation contexts in India. We analyze these capabilities in two villages, namely, Srikhola in Singalila and Rampuria in Senchal and also their social justice and dignity basis. We choose these villages as they are revenue villages and villagers have land rights. We also discuss Nussbaum's notion of thresholds.

Following this we evaluate Srikhola and Rampuria for how its families fare according to the three capabilities. Nussbaum's capabilities list is a proposal. It needs context, both political and geographical. Political contexts are forest rights, the Forest Rights Act, 2006 (FRA), and capabilities thresholds available in constitutional Fundamental Duties and Fundamental Rights. Protected Areas are the geographical context. We ask in Srikhola and Rampuria, what practical reasoning, affiliation and material and political environment control capabilities prevail? We use the FRA and Fundamental Rights as thresholds and assess their level in these two forest-edge villages. We conclude with comparative observations.

Capabilities, dignity and social justice

The three aforementioned capabilities as per Nussbaum's list are defined next.

> *Practical reason.* Being able to form a conception of the good and to engage in critical reflection about the planning of one's own life. (This entails protection for liberty of conscience.)
>
> *Affiliation.* Being able to live for and in relation to others, to recognize and show concern for other human beings, to engage in various forms of social interaction; being able to imagine the situation of another and to have compassion for that situation; having the capability for both justice and friendship. Being able to be treated as a dignified being whose worth is equal to that of others.
>
> *Control over one's environment.* (A) *Political*: being able to participate effectively in political choices that govern one's life; having the rights of political participation, free speech and freedom of association.
>
> (B) *Material*: being able to hold property (both land and movable goods); having the right to seek employment on an equal basis with others.
>
> (Nussbaum, 2011, p. 34)

Dignity remains the guiding principle in Nussbaum's capability approach. She states that her approach is 'one that lies at the heart of Rawls' project . . . the idea of citizen as a free and dignified human being' (Nussbaum, 1999, p. 46). Dignity is an intuitive notion and that is unclear. However, Nussbaum sees all of its notions as interconnected, where notions gain clarity from each other. She considers the notion of respect as particularly important. The 'basic idea is that some living conditions deliver to people a life that is worthy of the human dignity that they possess and others do not' (Nussbaum, 2011, p. 30). Nussbaum's is a theory of 'fundamental political entitlements' and so it employs a specific capabilities list. But she refrains from proposing a comprehensive quality of life assessment. Her political liberalism prescinds her from proposing a comprehensive value account unlike Sen, who identifies capabilities as the pertinent evaluator space for comparing life quality assessment (Nussbaum, 2011, p. 19).

The notion of human rights can be understood in the following manner – the capabilities of human beings should not be allowed to fall below a specific threshold as long as nation-states and the global community at large are able to produce that minimum threshold for each and every person (Nussbaum, 1997). Nussbaum's basic social justice claim is that 'respect for human dignity' requires that citizens be placed above an ample (specified) threshold of capability, in all ten of those areas (Nussbaum, 2011, p. 36). Her social justice approach's basic question is 'what does a life worthy of human dignity require?' In principle, human rights are for everyone, a notion that is echoed in the Indian Constitution in the form of Fundamental Rights.

Capabilities on Nussbaum's list are abstract. She asks, who will specify them. 'For the most part, the answer is given by each nation's system of constitutional law, or its basic principles if it lacks a written constitution. There is room for nations to elaborate capabilities differently to some extent, given their different traditions and histories'. The idea of threshold is part conception of her list. Her approach 'is a partial theory of social justice: it does not purport to solve all distributional problems; it just specifies a rather ample social minimum' (Nussbaum, 2011, p. 40).

The Constitution of India contains specific provisions on environment protection and the preamble to it ensures dignity of the individual. In a stressful conservation context, while employing the aforesaid capabilities, the Fundamental Rights of Right to Equality, Right to Freedom and Right to Constitutional Remedies are most pertinent. Article 14 of the Constitution guarantees people the right to be equally protected by the country's laws. Article 15 of the Constitution guarantees social equality and equal access to public areas and states that the state shall not discriminate on the basis of religion, race, caste, sex or place or birth. Article 19 guarantees freedom of speech and expression and Article 21 gives a citizen the right to life. Articles 32 to 35 provide citizens the rights to move a court of law in case of denial of fundamental rights. In addition to these rights, the chapter on Fundamental Duties in the Constitution is relevant in the conservation context. Article 51-A(g) of the Constitution states, 'It shall be duty of every citizen of India to protect and improve the natural environment including forests, lakes, rivers and wild life and to have compassion for living creatures'.

In 2006, India passed the Scheduled Tribes and Other Traditional Forest Dwellers Act or the Forest Rights Act, a landmark legislation which is a precedent in itself. This Act recognizes rights of Scheduled Tribes and other forest dwellers to use, protect and manage forest resources (Forest Rights Act, 2006). It emphasizes four main rights of the traditional forest dwellers, namely, Title Rights (mainly land that they cultivate), Use Rights (to minor forest produce and grazing areas), Relief and Development Rights (rehabilitation in case of forced displacement or illegal eviction) and Forest Management Rights (protection of forest and wildlife) (Forest Rights Act, 2006).

The Darjeeling hills are a contested landscape with various groups staking their claims (Lama, 2008). Darjeeling has seen regular upheavals and shutdowns for a separate state that have often affected how people view and value their forests. The British settled many communities in forest fringes in the 1800s. But they have become traditional forest dwellers under the

2006 Act. However, the Act's provisions have not yet been implemented here. This alienation from the forest land could affect their ability to manage forests, including safeguarding them from vandalism during periods of political upheavals. The Forest Rights Act has as its basis the fundamental rights of equality and freedom.

Capabilities and thresholds in Srikhola and Rampuria

The aforementioned capabilities are indispensable for achieving sustainable development. In our study area (see Figure 3.1), through practical reasoning, people seem to have developed good observation prowess when it comes to the change of the environmental conditions and adaptations, among others. Through affiliation, people seemed to have developed compassion for one another. The third capability, namely, control over one's environment is used to assess how empowered people are to manage their surrounding environment through different types of institutional mechanisms.

Practical reasoning

Dwellers of forest-fringe villages are indispensable for the conservation of natural resources. Different types of treaties on ecosystem and biodiversity conservation have stressed upon the importance of incorporating the

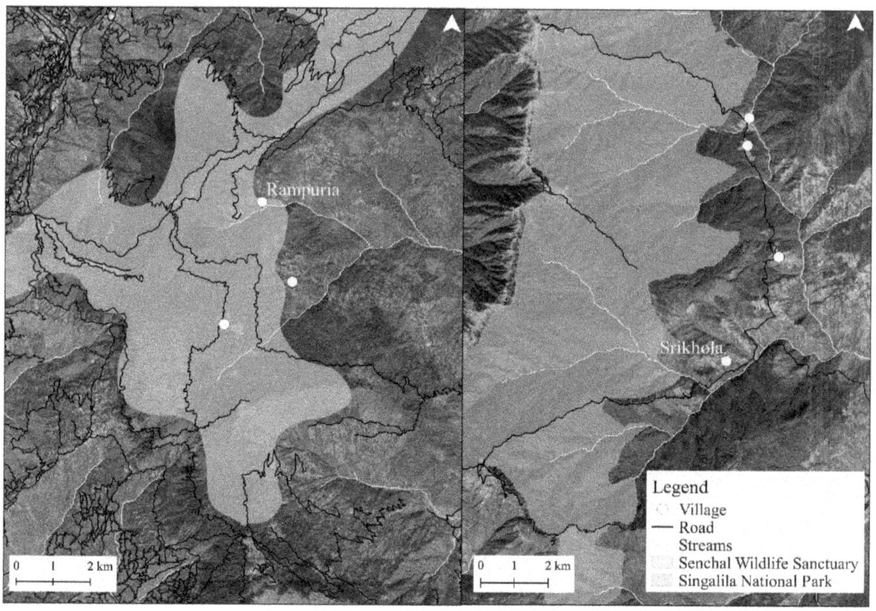

Figure 3.1 Srikhola and Rampuria: the topography

Source: Generated by Abhishek Samrat.

forest-fringe people in conservation policies. The observations, experiences and learning of these people is transmitted inter-generationally. Therefore, the rich knowledge that these people possess is often used for coping and adaptation in the face of different types of changes in the ecosystem. We have conceptualized the community's knowledge of the ecosystem and its change, their ability to differentiate between good and bad condition of the forest ecosystem with proper reasoning, adaption and coping procedure as practical reasoning capability of the community.

Observation and opinion of ecosystem services

CONDITION OF THE FOREST ECOSYSTEM

Interviews with villagers revealed how the perception of a good condition of a forest can be different for an ecologist and a protected area manager. The state refers to forests in terms of square kilometre of land under forested condition with a canopy cover of more than 10 percent. Thus, in this case, the more means the merrier. An ecologist would define it in terms of an ecosystem comprising a 'natural woodland unit' that includes all fauna, flora and abiotic elements, that is the more structural complexity the better.

On the other hand, people who depend on forests and live in close proximity to them articulate forest condition in terms of the fuelwood, water and edible wild fruits, flowers and vegetables that are depleting. Their measures are directly related to their ability or inability to access those resources. When forests degrade, so do those resources. Srikhola residents lament the degradation in the condition of the forest since 1992 owing to a forest fire followed by weed infestation. They also reported that the forest did not regenerate well due to the procedure undertaken. Moreover, some of the households from Srikhola blamed the choice of plant species for plantation activities. According to them, the Forest Department mainly planted pine trees, which is not indigenous and does not support biodiversity, nor does it provide much benefit for people's livelihoods. In the case of Rampuria, people think that the forest condition in terms of density has become better after it being notified as a Protected Area in 1992. However, there have been reports of illegal extraction, deforestation and increased dependency of a growing population that could alter conditions in an adverse way. A villager stated, 'Inside the jungle, edible wild vegetables are less as compared to before. Earlier we used to get the plant of khiraulo but now it has decreased very much. Now there are no cow sheds and we don't get bamboo shoots'. The cow sheds guarantee good manure but now there is paucity of food for the cattle. The same villager commented on flowers that have been destroyed. 'If they are eaten by the cows then it will grow much better the next year'.

People were asked what they regarded as a 'good forest'. Villagers responded from the perspectives of both development and conservation.

They said that forests should have a lot of trees and also that 'there should be enough amount of water, manure, and fuelwood for the villagers'. Interestingly, some of the villagers stated that the forest should be greener with lots of trees and wildlife. They further added that the forest should be able to provide enough food for the wildlife who would thus be less attracted to the farms for their food. Villagers support conservation of wildlife from their ethical and compassion point of view. However, some of the villagers also think the conservation of wildlife is related to maintaining balance in the forest ecosystem. These opinions and perceptions form the basis for sustainable development. These people have co-existed with forest ecosystems since the British settled in the region to clear forests and establish plantation units in the 1800s.

SOIL FERTILITY AND CROP PRODUCTIVITY

The forest has an indispensable role to play in maintaining the soil fertility through nutrient cycling and maintaining the moisture content of the soil through its hydrological services. Moreover, villagers collect dry litter from the forest, mix it with the excreta of domesticated animals and use it in their fields as organic manure. Additionally, being able to grow food and have access to healthy food is critical for food security. Our household interview responses suggest that the crop productivity has declined for both Srikhola and Rampuria. The degradation of the forest is the main reason for the decline of the soil fertility, as perceived by the villagers, since it is related to the decline of the soil moisture and reduced amount of leaf litter from the forest. Moreover, the number of livestock has also declined after the ban on grazing and access to the forest was restricted.

One of the villagers of Rampuria commented on previously mentioned situation: 'When the members of KPS [Krishi Prajukti Sahayak] had come last time they said that soil has become sour which lead to the growth of chhari amilo grass. The soil becoming less surely leads to the less production of crops'. He also threw light upon the decrease in rearing of cattle. 'There has been less number of houses rearing cattle where there is shortage of the availability of cow dung for the farm which leads to the low quality of the soil further leading to the less production of the crops'. Some of the people from Rampuria also felt that the decline in the soil fertility happened due to repeated cultivation of the same type of crop, year after year. Some villagers of Srikhola experienced a lot of soil erosion during monsoon. One of the respondents noted that the productivity initially increased after the use of chemical fertilizers, but after a certain time it started decreasing and presently, they get very little yield.

In both villages, the yield has declined drastically. A villager stated, 'Fertility has gone down and now we have stopped cultivating maize and ginger which is one of the most prominent cash crops. Instead new crops like, chillies, spinach, coriander is showing positive trend'. Yet another villager mentioned that they have stopped growing potatoes, millets, buckwheat and corn. In these circumstances, the role of the Agriculture Department can be very important as it is the main government institution responsible for all types of technology transfer related to farming. However, most of the villagers said that they do not get proper extension facilities from the Agriculture Department. A large section of them do not know about the village level worker (*Krishi Prajukti Sahayak*) of the Agriculture Department.

WATER PROBLEM

In both cases, villagers get water from small springs and streams. The people of the two villages concurred that the flow of water has declined in the streams and springs mainly due to the degradation of forest in the catchment areas. On one occasion, they had proactively raised some plantations in the catchment areas to increase the flow of water. The people observed, as the forest in the catchment areas degraded and there was weed infestation, the flow declined and the quality deteriorated. However, the people of Srikhola face severe water crisis especially during summer. When people were asked about the water problem, one of the villagers of Srikhola said, ‘There is absolute problem of water here’. In group discussions, people voiced the same thing. Rampuria villagers seem to be faring better than the Srikhola villagers. Their water scarcity issue is not as acute as it is in Singalila. However, some of the villagers of Rampuria pointed out from their observation that the frequent tremor in the Himalayan region is also related to drying of some of the springs and creation of new springs.

The water problem of Srikhola has manifested itself in another way, that is in the form of decline in soil moisture content which adversely affected the crop production especially in the upper part of Srikhola village. Very few villagers, who have their personal water sources, manage to irrigate their crop fields. Some of the households of Rampuria also informed us about the scarcity of water for irrigation purposes. However, some of them have adopted sprinkler methods of irrigation to adapt to this scarcity. The practical reasoning is also relevant to water resources to see how villagers differentiate between good water and bad water; and also how they use these resources. In most cases, people do not use water from the springs where there is pollution in the upstream areas. Therefore, people mainly use water coming from the forest and remain very selective in using water of the springs located inside the village. People in this hilly region generally boil water before drinking, therefore minimizing the chance of bacterial infection. Some of the households also observed the water scarcity problem and put overhead tanks in their houses so that they could harvest and use water efficiently. A group of people in Srikhola also planned to change their water sources when they observed the degradation of the quality and quantity, and approached the *Gram Panchayat* and changed the source.

HUMAN WILDLIFE CONFLICT (HWC) MITIGATION AND ADAPTATION

Community observations of the HWC trends, and their adaptation and mitigation measures, are some relevant and important indicators of the practical reasoning capability. Both of these villages are intensely affected by crop raids and livestock depredation. According to the people, there are several reasons for increase of HWC. First, Protected Area notification has caused the reduction pressure from different types of human activities like livestock grazing, poaching, felling trees which directly and indirectly helped in the increase of wildlife numbers. Second, wildlife translocation by the Forest Department has intensified this problem in several ways especially in the Senchal landscape. People have observed that the translocated monkeys are more comfortable raiding crops and snatching food in human settlements than in collecting wild foods. Third, the Forest Department chooses plant species for plantation which do not provide enough amounts of natural food for the wildlife, therefore wildlife faces food scarcity inside the forest and becomes more attracted to the farm. Fourth, people get very little support from the government for crop loss and livestock depredation which increases their problem all the more. The fencing provided by the Forest Department is not a good enough solution to protect wildlife like boars, bears and porcupines.

People of Srikhola have adopted different measures like beating drums, being vigilant of the animals and informing others. The Rampuria villagers on the other hand have converted a big portion of their agricultural land into private forest and use crop rotation, strong fences and strong cages for guarding livestock in addition to guarding their fields during the night. There is a community-wide opinion that the authorities have not addressed the problems of crop loss by wildlife depredation adequately. This is being addressed as these communities work with local non-profit organizations to set up barriers of movement that will not adversely affect the wildlife but provide farmers with some kind of crop assurance (Parvaiz, 2017). A lot of villagers in both Srikhola and Rampuria have diversified their livelihood activities with some non-farming activities to adapt to the HWC. The Forest Department has permitted them to use electric wire to prevent HWC, but they do not prefer to use it much due to the risk of electrocution.

ENERGY

In the Darjeeling Himalayas, fuelwood is indispensable for the villagers, not only for cooking, but also for space heating, boiling water and drying cloths during the monsoon season. Almost all the villagers of Srikhola depend on fuelwood availability and access for heating and cooking purposes. The communities living in close proximity to Protected Areas are highly dependent on firewood even though some find liquid petroleum gas (LPG) convenient for cooking purposes. The lack of proper road connectivity makes it difficult for villagers to transport cylinders to their villages in

Singalila and hence makes them entirely dependent on the forests. On the other hand, more households in Rampuria use both LPG and fuelwood to meet their energy requirements. Better road connectivity eases access and reduces constraints on transport. The villagers observe that the availability of fuelwood has decreased for both these villages. However, the condition of Srikhola is more severe than that of Rampuria due to the degradation of the forest. Nowadays, people have to travel long distances for the collection of fuelwood. Additionally, the fuelwood decline is also linked to the ban on tree felling.

These situations have practically meant that people plant saplings inside the forest with the support of the Forest Department and also become open to new technology such as the improved cook stove. Village-level self-help groups have been trained in building the improved stove, that has been found to be fuel-efficient, cause less pollution inside the house and enhance welfare among people (Barnes et al., 1994). Some of the people have also adapted their fuel-use practice accordingly. Households having access to LPG use fuelwood for boiling water, space heating, drying cloths whereas use LPG for cooking only. The improved cooking stove has been found not adequate for space heating purposes. Therefore, some of the households keep both the improved cooking stove and traditional cooking stove. They use the traditional cooking stove during winter only for space heating.

SUSTAINABLE MANAGEMENT OF THE FOREST

When people were asked what they do for the sustainable management of the forest, they provided positive replies and gave different types of methods in which they engage. For instance, they do not harm plants or trees during the collection of forest products, do not cut trees and encourage more plantation, among other things. A lot of people criticized the the forest management choices. For example, people of both the villages think that local indigenous species should be used for plantation activities which can enrich biodiversity and provide natural food for the wildlife. They expect the Forest Department to implement specific plans for weed management.

People feel that the wildlife menace can be managed in a better way. This by stopping improper relocation of animals, specifically in the case of monkeys in the Senchal landscape, and also selective killing of the hazardous wildlife species. People, especially in Srikhola, remain without hope about their participation in forest management activities. A lot of villagers have started tourism activities as an alternative to the decrease in yield from farm and forest.

DEVELOPMENT ASPIRATION

There is a general tendency among the villagers of the Darjeeling Himalayan areas to leave farming and engage in non-farming activities. The preferred form of this is being a daily-wage labourer in nearby towns and cities. As one farmer shared:

> By staying here we cannot do anything. . . . I want to impart my kids with good education. I don't think that they will do cultivation as they are not interested in agriculture so I want to provide them with good education so that they can get jobs.

Due to the problems associated with farming, human-wildlife conflict, declining yield from agriculture and uncertainty in rainfall pattern resulting in non-assured return, people have lost interest in farming and are hence pursuing other non-farming activities. Another respondent stated:

> We don't want a lot of wealth . . . our children are our wealth . . . we want them to do well . . . let them study . . . we'll do whatever we can . . . we don't mind taking loans to educate them. We want them to become *thulo manche* (prosperous and happy).

While education is valued, agriculture is no longer welcome as something to pursue. A respondent stated, 'I don't want them to do agriculture. Well because our father did it, we are doing it . . . now if our children also do the same thing then I realize it is a little difficult. Times have changed'. They also want their children to do non-farming activities in the future so that they do not have to face the problem that their parents are facing. Most of the villagers of Rampuria and Srikhola echo this sentiment. However, some of the villagers from both the villages think that their children should study and be educated. Then they should come back to their village and make the village a better place to reside, through hard work.

OPINION ABOUT TOURISM

Most of the people gave positive responses about tourism. First of all, people earn by arranging tours. Tourism also creates a positive impact on the economy of the village. They reported that they can learn a lot of new things from the tourists. However, a lot of people especially from Rampuria revealed that the tourists make the place dirty; therefore tourism is not exactly good from an environmental point of view. Srikhola village is situated on the trekking route of Sandakpu, a favourite tourist destination. A number of villagers have started home-stays there. Many tour guides are trained in sustainable tourism. They guide the tourists on how to keep the environment clean. On the other hand, not many tourists come to Rampuria. However, a number of villagers from both villages want tourism to develop in their respective areas in the future. People in general do not perceive tourism having a negative impact on the rural culture or they perceive it to be minimal.

Affiliation

An important capability for sustainable development is the capability of affiliation. This involves having compassion for others in crises like landslides, earthquakes, deaths, human-wildlife conflicts, forest fires and water scarcity. Also actively helping people to cope with different types of disasters. There are numerous examples of this capability in both regions. For instance, villagers in Srikhola and Rampuria are sympathetic towards the people who are disabled, often physically transporting them on backs when the emergency arises. The respondents opine that women face a lot of hardship especially while collecting fuelwood, carrying water to mitigate the water problem of the whole community and extracting non-timber forest produce owing to restrictions imposed by the Forest Department.

The role of the Samaj

'The *Samaj* is a vehicle for change and modernization on a community-wide basis. The *Samaj*, an indigenous social form, is important to observe as models of cooperation and community' (Messerschmidt, 1981, p. 42). This institutionalization of actions and help through the *Samaj* provides communities with a sense of security and belonging. There are rules and codes of conduct on how to assist people during times of crises. The *Samaj* is an essential feature of the hill communities that has survived the colonial rule and creation of the state of India. When a household is part of a *Samaj* it is mandatory for a member of the family to attend all meetings, gatherings, weddings in the village, deaths in the village or any other events of crises. The role of the *Samaj* is restricted to helping people out during crises and deaths in Rampuria. However, in case of Srikhola, the role also includes planting saplings and cleaning roads and drains to control water flow during the monsoons and avoid fatal landslides.

We assessed the individual's ability to think for the community. Almost all the villagers cited similar problems in their village, namely, water scarcity, lack of proper health facilities, lack of proper and motorable roads and absence of electricity. One of the respondents, when asked about the main problems of Srikhola, said, 'There is a problem of dispensary, road, water, etc. Most of the people use their own pipes for water'. Another person said, 'There is absolute problem of water here'. People have even questioned the role of the government body on the implementation of works which may be harmful for the community in the future. Villagers said that during the dry season, water is equally shared among community members by allocating a fixed time for water pipes to divert water into different households. One of the respondents of Rampuria said, 'There are developmental works that are left open with water flowing and I feel these could create landslides in the future'. Same was the opinion of one of the respondents of Srikhola when he responded about the faulty implementation of government work.

He said, 'If there is much damage then we do complaint'. People are also sympathetic about the environmental hardship that women have to face due to environmental degradation and restricted access. One of the respondents of Srikhola noted, 'They face many problems'. Some of the respondents of Rampuria reported that household chores like collection of water and fodder became very difficult for women in the face of environmental changes and that they consider it harmful.

Control over one's environment

This is one of the most critical capabilities and is very relevant to sustainable development in India. In fact it is an architectonic or meta-capability in India where paucity of resource access and control rights thwart flourishing. There are two sides to this capability – having control over one's material environment and having control over one's political environment that is being able to hold property and participate politically, respectively. This is a crucial freedom and one that is often neglected. We have conceptualized the environment of a forest-fringe dweller as the village in which she lives and the surrounding forest as the environment. The ownership of the land in the village and the surrounding forest can be seen as the material environment. The well-being of the forest-fringe villagers depends upon two main departments, namely, the Forest Department and the *Panchayat* and Rural Development Department. There are other line departments which work in those villages in coordination with these two departments. There have been lot of studies that indicate the importance of institutional mechanisms of conservation institutions such as Eco-Development Committees (EDC) and Forest Protection Committees (FPC) and development institutions like the *Gram Panchayat* for sustainable development (Behera, 2009). The ability to manage the village and the surrounding forest, the functioning of different types of institutions related to village development and forest management and the ability of participation of the villagers in those institutions can be conceptualized as the political environment.

Control over material environment

LAND RIGHTS

Since both Srikhola and Rampuria are revenue villages, the people claim to have land rights so that they can get credit from the banks as loans for agriculture, allied activities like livestock rearing and other small businesses by using the land as collateral security. Therefore, generally these villagers are economically in better positions when compared to the residents of the forest villages. The possession of land entitlement also makes them less vulnerable than that of the inhabitants of the forest villagers since they are not afraid of any kind of forced relocation.

FOREST RIGHTS

Since both the villages are revenue villages, and FRA had not been implemented in Darjeeling till the time we conducted our fieldwork, people do not have a right to the forest. The grazing of livestock and keeping cattle sheds inside the forest was banned in 1992 when it was notified as a protected area. None of the villages have any common grazing land. However, the Forest Department in both cases allows the villagers to collect small amounts of forest-provisioned services like fodder, fuelwood and non-timber forest produce. In any way, the collection of forest produce is not legally authorized. They are not allowed to cut trees. There are few cases of conflict between the Forest Department and villagers. One of the villagers of Srikhola said, 'Whichever guard sees us they chase us away. If they would only chase then it wouldn't have been a problem, they charge us fine as well'. While one aspect of the material environment threshold of owning land is met, the threshold performance in the other aspect of rights to forest is not satisfactory.

Control over political environment

The access of the forest to the people is more secure in the case of Rampuria, since they are part of the EDC. However, one of the respondents reported that there is access to forest but in stealth. He said, 'If there is news of forest guards being on duty then we don't enter the forest'. A number of families in both Srikhola and Rampuria have converted a part of their farm land into private forest, thereby, making it possible for them to meet a large part of their fuelwood and fodder needs from their private forest. Therefore, the constraints of accessing government forest do not hurt them much. But people without private forest access have to depend on the state forest and the problem of access restrictions to these forests hampers their well-being.

Forest rights does not only mean the right to forest resources, but also the entitlement of managing the forest condition for biodiversity along with sustainable flow of the ecosystem services. However, there is no provision of legal participation in the forest management in the case of the people of Srikhola. Villagers of Srikhola are very eager to form an FPC so that they can manage the forest in better ways. In the past, they proactively engaged in plantation in the forest land to check deforestation and soil erosion and maintain the flow of water. However, activity in the forest land has been largely restricted due to no entitlement of the residents of Srikhola.

The existing village-level institutions are co-operative societies, *Gram Panchayat*, self-help groups and different types of *Gram Samaj* within the village. People sometimes discuss the conservation of forest, need for plantation and adverse effects of deforestation during the meetings of these institutions. These discussions are very important for shaping individual behaviour for maintaining and improving the forest ecosystem. However, these institutions are not primarily entitled for managing the forest ecosystem and there are legal restrictions against their conservation activities inside the forest land. On the other hand, half of Rampuria comes under an EDC and the other half comes under the FPC.

However, a lot of people do not even know that there is an EDC and an FPC in their village. Most of them are not concerned about the role of these institutions. The majority of the people who participate in the meetings of the EDC do not speak and there is a centralization of power within the EDC. One of the respondents said, 'EDC is there and I am a member however we don't say anything. The literate ones speak we only listen'. Some of the respondents opined that there are discussions about conservation of the forest in the meetings of EDC and FPC. EDC mainly works on plantation and building of drains. People are less concerned about the activities of the FPC. One of them said, 'There is no EDC in the village. There may be an FPC because during the meeting they were talking about conserving animals'. Another person said, 'There are no EDCs or FPCs. No one holds awareness programmes about the environment'. However, both males and females participate in the EDC. The role of other village-level institutions like *Gram Samaj*, *Gram Panchayat* and SHGs is negligible.

DECENTRALIZED DEVELOPMENT

The *Gram Panchayat* is the main developmental institution in the rural areas. However, in the Darjeeling Himalayan region there was no *Panchayat* election from 2001 onwards due to the *Gorkhaland* agitation. The *Gram Panchayat* functions in a defunct mode with very limited roles and responsibilities for village development. In most cases, hopes and aspirations of the people cannot come to the surface. These two villages also face the same problem. The *Gram Panchayat* mainly functions as the agency for providing certificates and implementation of wage employment under MGNREGS. Different types of government programmes are implemented in a very poor manner. These include agricultural development, livestock development and other employment-generating activities. Decentralization is supposed to guarantee increased political participation which then translates into control over political environment capability.

There are other active village-level institutions in these two villages like *Gram Samaj*, self-help groups, EDC and FPC. *Gram Samaj* mainly functions in events such as marriages, deaths, natural calamities and is an institution through which people help each other during the time of distress. Sometimes, *Gram Samaj* also implements some conservation activities like plantations. EDC in Rampuria is also involved in developmental activities in the village through distribution of housing facilities to selected beneficiaries. People's participation in *Gram Samaj* and SHGs is better than in government institutions.

In the landscape of Darjeeling Himalayas, there are numerous examples where communities have applied to or petitioned higher revenue and forest authorities for bringing in facilities like a dispensary, electricity, road connection in their respective villages. The villagers of Rampuria have applied for roads and electricity and have received it. The condition of Srikhola is comparatively worse off. Facilities like road and electricity have yet to reach the village.

Conclusion

Nussbaum's theory of justice, as stated before, is substantive but not a complete theory of social justice. It is not limited to the notion of political justice or to liberal democracies but to all human beings. This is regardless of whether they are living in a liberal democratic state or are severely disabled. Her theory focuses on thresholds that nations should guarantee their citizens, but it does not address the question as to what social justice means once these thresholds are met. This raises the question of theory of justice specifying not just rights but also duties.

However, the question of whose responsibility should it be when it comes to realization and the expansion of the capabilities remains unanswered. Once institutions are set up for the delivery of justice, the question arises as to what the duties of individuals are in this scenario. Is it their duty to merely comply with institutional rules? Or do they have additional duties of promoting social justice?

There is no straightforward answer to these questions, specifically in the case of stressful conservation situations and fast-changing landscapes like those of Srikhola and Rampuria. It is difficult to zero in on which duties should exclusively be of the villagers in these two regions. Resilient as they may be, it is unfair to put the onus of social justice in this conservation context on these villagers. The answer lies in increased ways in which they can exercise their agency and take decisions that they value, rather than following a fixed predetermined path of justice in which they have little or no say. How does one then make this process more pragmatic? Participation in a socially just rule-making process is key as it is an important condition of democracy (Holland, 2014). As famed Roman poet Ovid states, 'the end doesn't justify the means'. It is thus important to increase focus on procedural justice in order to contribute to a more comprehensive theory of the same.

So, how do Srikhola and Rampuria fare in the social justice and capabilities department? While their material environment control and affiliation is at threshold levels, the same cannot be said for their political environment control. Nussbaum's framework seems to suggest that they do not fare well overall. The thresholds suggested by her in her framework are the bare minimum a government can provide a citizen. 'Justice in the life and conduct of the state is possible only as first it resides in the hearts and souls of the citizens', said Plato. However, residents in Srikhola and Rampuria report struggling with basics such as water, electricity and dispensaries. Human-wildlife conflict is also rampant and the political will of the state here does not seem to coincide with the well-being of the people, specifically what they regard as well-being. This is a problem with governance in Himalayan spaces such as these. While there are decentralized organizational arrangements, they are inadequate. They have rights but their political participation is limited and, in some instances, hindered. How will then the state reside in the heart and soul of its citizens?

The protection of the environment is a fundamental duty of every Indian citizen and is a guiding principle in Article 51-A(g) of the Constitution of India. Residents of both Srikhola and Rampuria, however, lack the privilege of improving their natural environment let alone live in them. The deprivations of capabilities also translate in the deprivation of their Fundamental Rights. Despite meeting threshold levels, there are capability deprivations. The lives of these residents have been structurally marginalized and they are deprived in terms of capabilities and more so, these deprivations are not their doing.

Article 48-A of the Constitution says that 'the state shall endeavour to protect and improve the environment and to safeguard the forests and wildlife of the country'. The forests can be safeguarded, sure, but what about the forest dwellers and dependents? As per Article 21 of the Constitution, 'no person shall be deprived of his life or personal liberty except according to procedure established by the law'. That is to say that everybody has a right to life as do the residents of Srikhola and Rampuria. There is the law but they are not the recipients of justice. They, as Nussbaum suggests in her framework, have little control over their environment which translates into a severe capability deprivation. In this case, Rampuria fares better than Srikhola mainly because it has an EDC.

In the affiliation aspect, overall, people fare well. They seem to have compassion for each other and organize themselves to help each other during crises. However, coping with crises is not the same for everybody and homogenizing these experiences is not fair. As stated before, women face the brunt of any crises because their share of labour is more and they taste less fruits of this labour. When it comes to practical reasoning, communities in both regions are resilient and have coping mechanisms in the face of change in the form of tourism. Both regions are ready to embrace tourism and the boost in the economy that in their opinion it will bring. Rampuria is, however, disadvantaged because not many tourists come there.

Environmental equity is a vital component of social justice. When it comes to social justice, Nussbaum's focus on the individual becomes an important factor for analysing society in both regions. As mentioned before, the recipients of justice cannot be all the same. A woman forest dweller has different capability deprivations than a man, a child is more vulnerable than others and so on. The poorest and most marginalized sections of the community are often overlooked and neglected. There is a need to theorize justice intersectionally while assessing forests, which is as important as simply listing down laws and rights that do not apply in the same way to all people.

In a capability context, expansion of capabilities cannot be solely an individual enterprise. While assessing individual capabilities is important, it is also essential to shift the focus of capabilities analyses from the individual to a collective. There is hence a dire need to examine the importance of social structures and the interaction between individual capabilities and social structures, the structural aspects being families, government institutions and other social institutions like *sabhas*. This brings us to the generation of collective capabilities which can be created through a process of collective action. Also, the collectivity at large and not just a single individual can be beneficiaries of these new capabilities (Ibrahim, 2006). The goals of justice here are hence not merely focused on the individual or a community but on constant navigation between the two. Thus, if the collective is able to organize itself and existing institutions aid in the same, through collective agency, the group can gain new collective capabilities (Behera, 2009). These capabilities then in addition to the individual capabilities would benefit both the individual and the collective while nurturing social capital which can aid in emergence of context-specific solutions to complex conservation problems.

References

Barnes, D., Openshaw, K., Smith, K. and van der Plas, R. (1994). *What Makes People Cook with Improved Biomass Stoves. A Comparative International Review of Stove Programs*. Energy Series. World Bank Technical Paper, 242.

Behera, B. (2009). Explaining the Performance of State – Community Joint Forest Management in India. *Ecological Economics*, 69(1), pp. 177–185.

Comiling, K. (2014). A Postcolonial critique of Amartya Sen's Capability Framework. *Perspectives in the Arts and Humanities Asia*, pp. 4–15.

Holland, B (2014). *Allocating the Earth: A Distributional Approach to Protecting Capabilities in Environmental Law and Policy*. Oxford University Press.

Ibrahim, S. (2006). From Individual to Collective Capabilities: The Capability Approach as a Conceptual Framework for Self-help. *Journal of Human Development*, 7(3), pp. 397–416.

Lama, B.B. (2008). *The Story of Darjeeling*. Darjeeling: NYL Publications.

Messerschmidt, D. (1981). Nogar and Other Traditional Forms of Cooperation in Nepal: Significance for Development. *Human Organization*, 40(1), pp. 40–47.

Nussbaum, M. (1997). Capabilities and Human Rights. *Fordham Law Review*, 66, p. 273.

Nussbaum, M. (1999). *Sex and Social Justice*. Oxford: Oxford University Press, Reprint, 2013.

Nussbaum, M. (2000). *Women and Human Development: The Capabilities Approach*. Cambridge and New York: Cambridge University Press.

Nussbaum, M. (2011). *Creating Capabilities: The Human Development Approach*. Cambridge, MA: The Belknap Press of Harvard University Press.

Parvaiz, A. (2017). *Darjeeling's Small Farmers are Losing 40% of Their Crop to Raiding Wild Animals*. [online] Scroll.in. Available at: https://scroll.in/article/832715/darjeelings-small-farmers-are-losing-40-of-their-crop-to-raiding-wild-animals Last accessed on 2 March 2020

4 Sustainable ecological capacity of forest-fringe families

Soubadra Devy, Annesha Chowdhury, Abhishek Samrat and Sarala Khaling

Introduction

Forests sustain human communities through ecosystem goods and services. They are the source of fuelwood, fodder, medicinal plants and clean and safe drinking water. Forests support agriculture through pollination and nutrient cycling. Forests are also spiritual and recreation sources for different cultures (Daily, 1997; Millennium Ecosystem Assessment, 2005; Menon et al., 2009). When people have access to ecosystem goods and services, it contributes to their well-being (Daily, 1997; Costanza, 1997; Millennium Ecosystem Assessment, 2005). India is home to 708,273 square kilometres of forest which accounts for 21.54 percent of its land cover (India State Forest Report, 2017). Approximately 275 million people in India depend on forests (Lynch and Talbott, 1995).

However, forests in India are state controlled where access varies according to the level of protection (Borrini et al., 2004; Karanth and Nepal, 2012; Lele and Menon, 2014; Kothari, 2013). The state curtails access rights to traditional forest resources (Brockington, 2004; Saberwal, 1999). Blanket policies protect gregarious species, like wild boars that raid crops. Besides, protection has led to the increase of predators like leopards which has led to human-wildlife conflicts (Mishra, 1997). The most controversial outcomes of these policies are the large-scale evictions of people from PAs (Torri, 2011; Rangarajan and Shahabuddin, 2006). Protected Area (PA) laws sever access to the forest goods and services that sustain communities thus affecting well-being (Bhullar, 2008; Sarin, 1998). Conflicts and negative relationships between the state, conservation actors and forest-fringe communities abound (Ogra and Badola, 2008). The legislation of Forest Rights Act (FRA) of 2006 is meant to entitle and safeguard traditional forest dwellers. Both individual land rights and community forest rights of the FRA ensure well-being.

The meaning of human well-being has evolved from maximizing human pleasure (Kovac, 2012) to that of raising 'eudaimonia or human flourishing' (Ransome, 2010, p. 44). Income and health are parameters to assess a citizen's quality of life or well-being (Millennium Ecosystem Assessment, 2005).

People's ability to derive benefits from their surrounding ecosystem contributes to their well-being (Daily, 1997; Sandhu and Sandhu, 2014). Current approaches to well-being borrow from Aristotle's concept (Sen, 1999, 2009; Nussbaum, 2011; Ransome, 2010) of 'human flourishing' (Ransome, 2010, p. 44). These prescribe a more rights-based 'plural' approach (Ransome, 2010, p. 41), which emphasizes 'freedom of choice' (Ransome, 2010, p. 41). A meaningful measure of well-being is a true representation of the quality of life of any individual (Nussbaum, 2011). The conditions of well-being are a political entitlement (Nussbaum, 1997). Income, education and a healthy environment are necessary conditions to achieve a minimum level of well-being (Sandhu and Sandhu, 2014) and form the 'meta-capability' of an individual (Holland, 2008b, p. 321).

Sustainable ecological capacity of forest-fringe communities

Nussbaum's (2011) and Holland's (2008a,b) expansion of Sen's (1999) theory on *human capabilities* were developed as a theory of well-being. This framework was advanced to provide an 'evaluative space' and a 'political philosophy' to assess justice and freedom. It was to emphasize the importance of 'open debate' about 'just values and principles' in addition to the Rawlsian 'just institutions' to achieve social justice (The Economist, 2011). Capabilities are political entitlements that the state and society must protect for well-being. Nussbaum lists ten facets that affect human life as 'central capabilities' to articulate what well-being measures should include. These are rights or political entitlements the protection of which is the moral obligation of the state and society. Holland illustrates that the conditions on which the central capabilities depend are achievable through a 'meta-capability' – one that could enhance or impede the achievement of all other capabilities. Holland introduces the term 'Sustainable Ecological Capacity' (SEC) as a 'meta-capability' for well-being along with income and education.

'SEC' was developed as a 'guiding principle' for how the world ought to distribute environmental liabilities and benefits. Nussbaum's theory of justice, in this regard, was found to be restrictive in nature because 'the natural environment's instrumental value only explicitly figures into the eighth capability on Nussbaum's list of central capabilities' (Holland, 2008b, p. 322) and doesn't necessarily capture the critical role that ecosystems and environments play in human societies. The same constraints were found in the economic frameworks for evaluation of environmental policies, which have been criticized for being too utilitarian and hence unable to capture the instrumental role of the natural environment in people's lives (Holland, 2014). Given the limitation of Nussbaum's 'partial theory of justice', Holland proposes the concept of SEC as a meta-capability that supports all the central capabilities listed by Nussbaum. Holland emphasizes that a healthy ecological functioning within people's natural ecosystems on which they are especially dependent is a key condition or requirement for individuals to realize other capabilities in order to achieve a minimum level of well-being because these are the conditions that 'enable the very possibility of human life' (Holland, 2008b, p. 324). By this argument, this puts SEC at par with other conditions of well-being such as owning property or education thus becoming that individual's 'opportunity set' – key conditions that make it possible for individuals to experience and live a meaningful life, with dignity but also recognizing that the ability of humans to be able to exercise the central capabilities is sometimes possible independent of social, political and economic contexts; however, that is not the case with the natural environment (Holland, 2008b, p. 320).

She elucidates that the ways in which social, political and cultural human institutions define and mould people's ability to convert resources into valuable accomplishments, elements of ecological systems on which people depend, influence the conversion of ecological functioning to valuable goods and services in the same manner (Holland, 2008b). This means that desirable levels of functioning in the ecosystem such as tree regeneration, soil nutrient cycling or seed dispersal, waste recycling and water purification have an important influence over the ability of people to harness and convert that function to useful, tangible benefits such as fuelwood, fodder, healthy soil organic matter, a clean and safe environment and drinking water respectively. The statement is supported with two conditions that are key for the SEC to be achieved by an individual. These include the protection of ecological systems in a manner that continues to support life on earth and continue to the sustained flow of critical conditions that 'enable threshold levels of other central human functional capabilities' (Holland, 2008b, pp. 325). Holland ends the argument by stating the responsibility of protecting the ecological systems and ecosystem processes that people depend on with the government that 'promises to sustain the conditions of life on earth' for the achievement of capabilities (Holland, 2008b, p. 325).

In this chapter, we present empirical evidence that access to a healthy, functioning ecosystem is an important condition to achieving well-being. Additionally, we argue that single well-being conditions such as income and wealth do not provide the entire context in which an individual's well-being can be understood. In fact, it can be misleading. Forest communities flourish when they meet certain ecological, social and political conditions. We use an interdisciplinary approach to empirically assess SEC or 'meta-capability' of a forest-fringe community. We operationalize SEC as a set of social and ecological conditions that are an individual's 'opportunity set'. We then use our findings to critique Holland's theory and suggest a more expansive framework for SEC.

Forest, revenue and plantation villages

We assessed the SEC of households in seven villages in close proximity to the Singalila National Park and Senchal Wildlife Sanctuary (see Figure 4.1).

Figure 4.1 Singalila and Senchal: demarcating forests and villages

Source: Generated by Abhishek Samrat.

The national park is under a stricter protection regime in comparison to the wildlife sanctuary with relatively more access to the community. We selected study villages based on their location near the Protected Areas. Their forest dependence, market connectivity and income diversity were used as criteria for selection. The villages had access to forests and were willing to voice their grievances. But their access levels differed. This difference has to do with access abilities and access has to do with jurisdiction. The seven villages were selected from three dominant jurisdiction types in the region. A description is provided for reference. 'Forest Villages' were established by the Forest Department for landless forest labourer (Sharma and Sarma, 2014, p. 42).

In many parts of India, currently, such villages suffer as 'illegal' settlements or 'encroachments'. Their use and access to forest resources remain restricted. The irony is the colonial state created these villages for logging timber and establishing plantations. The 'Revenue Villages' fall outside of Protected Areas and Reserved Forests. Communities here have land rights. Their main income is from agriculture. But they depend on the Forest Department and the *Panchayat* for forest access permits. By 'Plantation Villages' we mean rural units established to maintain Cinchona tree plantations. They are managed by the state. We exclude tea plantation units from this evaluation as they are beyond the scope of this study. Cinchona plantation villages have landless people with restricted forest access. Surprisingly, the village we selected neither fell within the purview of the Forest Department nor the Cinchona plantation. Therefore, their situation to exist as a village is precarious. Residents of this particular village are not eligible for development schemes by the Forest and Revenue Department.

Historical context and social condition

Forest management and regulation in India started as early as 1835, however it was the forest policy in 1855 which gave unequivocal power to the government to designate tree-covered land as state property and made no provisions for user rights (Kulkarni, 1987). Pre-independence Darjeeling was always managed as a special unit by the British, first as a 'non-regulated area', then as a 'Scheduled District', followed by designating it as 'Backward Tracts' and then as a 'Partially Excluded Area' until independence (Lama, 2008, p. 167). This was to separate the hill tribes from the mainstream national movement and to ensure security of the frontiers. The British acquired Darjeeling from the *Sikkim Chyogyal* or King as a sanatorium for their officers (Besky, 2008). Darjeeling was then vegetated with wet montane forests and moist deciduous forests (Gamble, 1896). The men of the colonial era sought to transform this landscape into a 'romantic getaway'. The British settled workers from neighbouring areas to clear these forests. Under the British *Raj*, forested landscapes were often designated 'wasteland' in order to overthrow the customary rights of communities and bring these lands under the purview of the erstwhile Forest Department for commercial gains.

Darjeeling, having had a history of contested ownerships proved to be the perfect candidate for commercial exploitation. They raised softwood and hardwood plantations (Banerjee et al., 2010). The British managed plantations for profit through scientific forestry. The Forest Department settled workers in small hamlets inside the forests. Workers gained access to every part of the forest while gaining an intimate knowledge of the forests, the trees and resources. These forests were also a source of subsistence for the settler community. They farmed for subsistence and reared livestock that were grazed in these forests. *Gaucharan* or *Gote* were the common forest areas where farmers put up cattle-sheds and allowed their livestock to graze. An elderly respondent from Gorkhey village reminisces: 'Until 30 years ago we used to pay 50 paisa per cow to rent space inside the forest for our cattle-sheds. The government has taken over the forest and removed all cattle-sheds and made it into a national park'. The abundance of manure in the past was cited by the respondents. 'Cows were free in the forest, they used to eat these bamboo shoots and defecate there so there was a lot of manure there which helped the tree saplings grow abundantly'. Rocks, certain trees and fresh water spring sources evolved into 'sacred' symbols. Soon, these forests became cultural spaces holding different meanings for forest communities.

Post-independence, the forests and the hamlets they contained came under the jurisdiction of the state, according to the Government of India Act, 1935 and the National Forest Policy of 1952 (Kulkarni, 1987). The state's need for the services of the forest communities gradually declined. Landless labourers lost user rights post-independence. In 1952, the Forest Department took control from the regional feudal system and established PAs (Hazra, 2002; Narain, 2015). The area under reserve forest increased. In the late 1980s and early 1990s, many reserved forests became notified and gained Protected Area status. The Forest Department labelled many forest communities as encroachers and relocated them. Villages at forest fringes found the same activities curbed and designated as 'illegal' (Narain, 2015).

Ecological condition

Change in forest cover in the context of Protected Areas from 1990 to 2015

Forests have undergone changes in structure and composition due to human interactions with them. Deforestation, invasive species, forest fires, lopping and understory removal have re-structured the ecosystem. Ecosystem processes and functions have thus altered. Yet, forests have also changed as a result of policies that have restricted entry of people. We assessed change in forest cover in three time periods, since PA formation in Darjeeling. We analysed high resolution satellite imagery of Singalila and Senchal for the years 1990, 2000 and 2015. These images include the village clusters in the proximity of these Protected Areas that were selected for the study. The remotely sensed years represent change before and after the setting up of the Protected Areas. We analyzed the images using standard remote sensing and GIS tools and metrics. The high-resolution satellite imagery was classified into old growth montane forests, recovering secondary forests, agricultural lands and abandoned cattle-sheds. Changes in the extent of each of these classes were compared across the three selected time periods. Differences between the two PAs showed up and we think are due to external drivers of change. These include Protected Area management policies, and factors such as road connectivity, proximity to markets and local bio-climate.

In 1990, Singalila forests were not yet designated as a national park. There was open access to forests, and grazing cattle inside the forests was not yet banned. People could collect their preferred fuelwood species from common grazing areas or *gote* inside forests. Satellite imagery from 1990 reveal more open, secondary or young forests than old growth wet montane forests. But in Senchal, already declared a wildlife sanctuary by 1986, forests were more homogenous. There was uniform distribution of biomass. Grazing was still allowed because the ban was effective only a couple of years later.

A decade later in 2000, Singalila's wet montane forest old growth cover increased as secondary forest decreased. This change can be attributed to the grazing and felling ban that followed the notification of Protected Areas in the 1990s. People abandoned grazing areas inside the forests. Secondary forest species colonized these areas. Respondents talk of a forest fire that gutted the forest and native *malingo* bamboo proliferated and established itself. A similar pattern followed in Senchal, but it was also marked by an increase in agriculture. The British had converted much of Senchal's forests into tea and Cinchona plantations. Here, optimal bio-climatic conditions for plantations and agriculture prevailed because of the average elevation (1,600 m to 2,100 m). Post-1990, there appears to be an increase in villages engaging in agricultural activities. Villages around Senchal had access to market and were successful in reaching it with their produce.

The year 2015 was chosen to represent and assess present conditions. In that year, the grazing and tree felling ban persisted on grazing and tree felling. We found a recovery in the burnt area by secondary forest in Singalila. There was increase of old growth wet temperate forest area inside the Protected Area. By 2015 we see a significant area of the Senchal landscape invaded by bamboo and open scrub. The Senchal landscape saw an increase in the construction of roads and housing.

A forest recovers but fuelwood availability declines

We used on-ground vegetation surveys to assess fuelwood availability and restocking in sections of forests where entry was not restricted. Singalila and Senchal's villages are dependent on their surrounding forests primarily for fuelwood and drinking water. We asked the communities to identify places where they collected fuelwood without violating Protected Area norms. We also asked them to identify areas which are well stocked and depleted of resources. This was done according to their perception of highest and lowest availability of fuelwood species. We also asked them to identify areas with minimal human influence that we could designate as reference forests and asked them to list trees according to their preference as fuelwood. We used this method to understand the condition of the forest not just in general forestry terms but in terms of the fuelwood species preferred vis-à-vis those that were available.

Families that were dependent on their surrounding ecosystem were selected for interview using standard methods from the seven villages of Singalila National Park and Senchal Wildlife Sanctuary. Our interviews reveal that all households in both the Protected Area clusters primarily depend on fuelwood to cook food, as cattle feed and for heat during the cold winter months. However, our surveys showed that 84 percent of the respondents in Singalila depend on the fuelwood-fired traditional *chullah* completely, whereas only 47 percent of the respondents from the Senchal landscape use the traditional stove only, with the rest compensating with liquefied petroleum gas. The amount of fuelwood consumed in a month varies anywhere between 440 pounds or 200 kilograms to 1,800 pounds or 816 kilograms.

Most fuelwood collection takes place between January and April in the form of 'dried up and fallen twigs and branches' (or *jhikra* as they are referred to in Nepali). While all of the respondents agreed that the availability of fuelwood had decreased considerably since their 'grandfather's time', the majority of respondents from the Senchal landscape attribute the decrease in fuelwood availability to the ban on cutting trees and the lack of older trees, most respondents of the Singalila landscape attribute it to the big forest fire or *darelo* of 1992, pressure from an increasing population of people collecting fuelwood from the forest and the ban on cutting trees due to the establishment of the PA.

Our assessments revealed that most fuelwood collection areas were within recovering or secondary forest zones. These are represented by a higher number of saplings compared to adult trees of the same species in both Singalila and Senchal. These areas lay outside PAs but were abundant with secondary species. Though they were plentiful, these species were not preferred by the village community. Communities in both Singalila and Senchal prefer the noble oak or *Quercus lamellose* and two other *Quercus* species that local people identified as *Buk* and *Phalant* (see Figure 4.2). According to some studies, this is a fuel-efficient wood (Rai et al., 2002). However, we did not encounter oak species in fuelwood collection sites of Singalila and Senchal.

Figure 4.2 Oak forest, Singalila

Source: Annesha Chowdhury.

We thus decided to further explore the 'non-availability' of the preferred fuelwood species. We compared the abundance and regenerating status of the preferred oak species to species that were not preferred but widely available (such as *Symplocos* spp.). We encountered a skewed ratio – more saplings than adults. This means fewer mature trees as compared to saplings. In Singalila and Senchal, the abundances of preferred fuelwood trees were very low compared to more abundant trees that are not generally preferred by the communities. This unavailability is further exacerbated by forest access restrictions as a result of Protected Area policies. Protected Area laws restrict forest access under the Wildlife Protection Act, 1972. Furthermore, to add to their woes, precipitous parts of the landscape remain inaccessible. While there is establishment of the bamboo species, it is now invasive and has invaded forest areas where oak species could have grown. The current unavailability of fuelwood was due to terrain constraint, oak's staggered ability to restock and bamboo invasion. But access restriction was also an important reason.

The political context

Relationships between people and their place have reversed. The core issue of indigenous political movements remains land rights (Watson, 2007). In 2006, India passed a progressive legislation – The Scheduled Tribes and Other Traditional Forest Dwellers Act or the Forest Rights Act. This act recognizes rights of Scheduled Tribes and other forest dwellers, 'to use, protect and manage forest resources' (The Economic Times, 2016). Countries like Canada, the United States, Brazil, Mexico, Bolivia and Australia recognize indigenous property rights. But such recognition is yet to happen in many parts of India (The Economic Times, 2016). The Darjeeling hills are a contested landscape with various groups staking their claims (Lama, 2008). Darjeeling has seen regular upheavals and shut-downs for a separate state that have often affected how people view and value their forests. The British settled many communities in forest fringes in the 1800s and now they have become traditional forest dwellers under the 2006 Act. However, the Act's provisions have not yet been implemented here. This alienation from the forest land could affect their ability to manage forests including safeguarding them from vandalism during periods of political upheavals.

Our household interviews revealed an interesting interplay between perception and reality. Our interviews show that there is a difference in how people understood the rules and provisions of their legal status as forest-fringe communities and the reality. All households in Singalila and Senchal villages have possession of the land they occupy; however, most respondents do not know the difference between having the security of tenure for that piece of land and having legal right to live in the land. Some don't have any knowledge regarding their tenurial status. In the Senchal landscape, 28 percent of the respondents claimed to have proper tenurial rights (*patta*) and said that entry inside the forest was allowed, 26 percent of the respondents claim they have land 'registration' but they actually have only occupancy rights and say that entry inside the forest is allowed.

Furthermore, 46 percent of the respondents of Senchal admit to not having any form of legal land rights, and also claim that entry inside the forest, especially the Protected Area, is not allowed. However, people do enter. All the respondents however were unanimous in their claim of being allowed to collect dry and fallen logs. On the one hand, in Singalila, the majority of respondents (66 percent) admitted to not having proper land rights and restricted access to forest, whereas around 18 percent of the respondents could not provide any information regarding the legal status of their land. One *Rai* resident of a village in Singalila commented on the present situation. He said, 'We have just like Samanden (papers for occupying the land) . . . but unlike *khasmal* villages, we cannot buy or sell this land. A person is allotted some land and has to use it to earn a living'. However, there is a small proportion of people (16 percent) who claim the papers that they have are actual land rights or the *patta*. This could also mean that there is a proportion of the population who may have traditional, indigenous property rights and owned land in the area before the British took over the landscape. Therefore, we ask what does this mean for the SEC of people living in the fringe-villages of Singalila and Senchal?

Resource ecology and resource access as a condition for SEC – a critique of Holland's definition

One of the principal goals of Holland's SEC was to critique Nussbaum and the existing policies of environmental justice because they did not fully incorporate the critical role of nature in supporting human life (Holland, 2008, 2014) and hence were unable to account for the disproportionate allocation of 'environmental benefits and burdens' especially among low-income, minority communities (Holland, 2014, p. 58). We begin with our first argument, that is our articulation of nature and nature's importance. By this we mean not just the natural environment that is instrumental to human life, but also the intrinsic or non-instrumental value that nature holds. Therefore, irrespective of the known role nature plays in human societies or the known 'values' that people ascribe to nature, the environmental justice worldview should move from considering people as mere beneficiaries of nature to rather, as part of it, co-producing, co-evolving as mutually interactive components of a larger system or more simply 'complex, integrated systems in which humans are a part of nature' (Berkes et al., 2008).

Second, to say that 'components of ecological systems can influence conversion of natural resources and goods into valuable achievements, the same way social, political and economic institutions shape an individual's ability to convert resources into valuable goods' (Holland, 2008, p. 324) is to ignore the control of human consciousness and power in making critical decisions while also unknowingly assigning 'agency' to ecosystems therefore exhibiting 'a tendency to attribute purpose and therefore value to nature and all of its processes as such' (Lele et al., 2013, p. 14). Therefore, in our study instead of the 'natural environment' as a whole, we use categories such as provisioning, regulating and cultural services, from the ecosystem services perspective (Millennium Ecosystem Assessment, 2005) to articulate the benefits, tangible and intangible, that people obtain from nature to measure the 'valuable achievements' (Holland, 2008b, p. 324). Here, we would like to take the opportunity to explain that benefits in the form of ecosystem services are distinct from ecosystem functioning and ecosystem processes and hence the categories represent all the final products that people are able to obtain from the ecosystem by using technology or effort to convert the 'product' of nature into a usable because people have assigned some kind of value to it (de Groot et al., 2002; Huntsinger and Oviedo, 2014). For example, a desirable level of ecosystem function could result in a well-wooded area. This could be either managed for timber or for carbon sequestration or for recreation benefits. The value is determined by who assigns it. This is how the same ecosystem can have multiple values and for different stakeholders. Which value gets importance and policy backing results in how the ecosystem is managed and therefore depends on 'who' are the stakeholders.

In the context of the people living in both Singalila National Park and Senchal Wildlife Sanctuary, devolution of user rights to the forests followed by the creation of Protected Areas led to the subsequent ban on grazing and cutting trees. There exists a restriction on fuelwood collection because of the villages' limited access to core areas which have the preferred fuelwood stock. Fuelwood is a key ecosystem good because of the sheer proportion of the demographic depending on it. Our study establishes that even with a recovering oak forest inside the protected area, fuelwood availability was dependent on villagers' access to parts of the forests where there was a healthy population of the preferred oak species, with a desirable number of mature/aging trees that could give branches and logs during intermittent weather conditions. The forests to which these people had access had more saplings than adult trees, even of the easily available but not preferred fuelwood. The restricted entry and the lowered restocking ability of the forests impeded the ability of these communities to be able to fully 'function' because of their negotiated capabilities of 'life and bodily health' and 'control over one's material and political environment'. The Senchal villages are able to compensate for this ecological status quo because of their growing connectivity and linkages to markets, which is not the case for Singalila. The latter still remains isolated from urban centres due to the lack of proper road connectivity. Senchal families overcome their energy constraint through easy access and use of alternate fuels such as liquefied petroleum gas. Their road connectivity and market proximity ensure access to the alternate fuels. But Singalila remains isolated and suffers from lack of road connectivity and market proximity. This provides an empirical support to previous studies suggesting ecosystem-based approaches to address poverty and justice for achieving well-being in the region (Sandhu and Sandhu, 2014).

'Sustainable Ecological Capacity is a meta-capability that enables all the capabilities on Nussbaum's list of central human capabilities. Having this meta-capability involves being 'able to live one's life in the context of ecological conditions that can provide environmental resources and services that enable the current generation's range of capabilities; to have these conditions now and in the future' (Holland, 2008b, p. 324). For this definition to hold, the presence of a well-stocked oak forest would be sufficient for maintaining this resource for 'now and in the future' and thus meet the conditions of the meta-capability. Our landscape ecology analysis, on the contrary, reveals that with the establishment of Protected Areas, the extent of original forests inside protection has increased. Also, our forest regeneration study helped us identify decreased ecological capacity of access areas. The availability of current fuelwood stock has declined due to restriction, terrain, ecology and use. This means 'to live one's life in the context of ecological conditions that can provide environmental resources and services . . . for now, and in the future' is not sufficient to encompass the SEC as a meta-capability of forest-fringe villages. For SEC to become a functional 'meta-capability', the conditions must include the concept of access and time. In order for the condition of well-being or SEC to fully support other capabilities, principles for distributive justice must be supported by principles of procedural justice. Thus, for SEC to be a holistic and useful idea or theory of environmental justice, it must include other aspects such as access and rights in addition to the ecological conditions that support that environment.

Holland further argues that the moral obligation of preserving the ecological condition lies with the state to ensure that future supply of nature's benefits that support people's capabilities are not irreversibly impaired. She further elucidates two conditions to achieve this goal: the first, protection of ecosystems 'to sustain conditions of life on earth' and second to protect ecosystems to sustain the flow of benefits (Holland, 2008b, p. 325). However, these recommendations do not resolve the contradictions that arise when applied in the context of communities living along Protected Areas, where the primary goal is protection of ecological conditions for the future. How does one, then protect the meta-capability of forest-dependent communities in an area that is already protected by the state but restricts their access? Our study revealed that the state has been successful in protecting the conditions of the forests by establishing Protected Areas. The forests are recovering and regenerating inside. However, the SEC of communities is compromised because of their high dependence on fuelwood, restricted access and access to only areas with high sapling numbers.

In order to address this dilemma, we adopt a more current discourse on conservation and social justice. We suggest that newer conservation models have adopted more participatory approaches, where the onus on 'preserving the ecological condition' of resources lies with the communities dependent on that resource. Such decentralized approaches complement the efforts of the state to preserve ecological conditions for the future while ensuring well-being for communities in the present (Palmer and Engel, 2007; Schreckenberg and Luttrell, 2009; Pelletier et al., 2016). For sustainable ecological capacity (SEC) of a household, we therefore suggest access to ecosystem goods and availability of goods to be used as meaningful indicators of evaluation of social justice in the context of environmental policies and decision-making. Here 'availability' also means ensuring the conditions for regeneration of forest areas that are devoid of oak tree populations. Through our understanding of the historical and socio-political context of this landscape, the implementation of the Forest Rights Act (FRA), 2006, could be better for the forest condition. People may view forests as state property with a sense of alienation without rights. However, implementing the FRA can ensure ownership and sustainable energy use incentives. Communities can also safeguard against large-scale defacement. The authorities, therefore, must ensure the ecological, social and political conditions for communities to achieve a threshold level of their SEC in order to enable them to achieve other capabilities of life, health, control over their environment, senses emotion and thought, affiliation and concern for other species to name some. The SEC along with income and education will then form the 'opportunity set' of people living in the fringe of Protected Areas.

We conclude that by adopting a non-economic and a more capabilities-oriented framework for evaluation of the Protected Area policy in the context of fringe families, we were able to develop measurable indices for operationalizing Sustainable Ecological Capacity as a working concept and not just a theory of environmental justice (Holland, 2008, 2014). By using standard methods in ecology and the social sciences, we are able to show that evaluative frameworks for environmental justice require measurable methods to successfully assess the effectiveness of a policy.

References

Banerjee, A., Ghosh, S., Springate-Baginski, O. and Nine, N.F. (2010). *Obstructed Access To Forest Justice in West Bengal State Violations in The Mis-Implementation Of The Forest Rights Act 2006*. Manchester: Institutions and Pro-Poor Growth (IPPG), University of Manchester.

Berkes, F., Colding, J. and Folke, C. (eds.). (2008). *Navigating Social-Ecological Systems: Building Resilience for Complexity and Change*. New York: Cambridge University Press.

Besky, S. (2008). Can A Plantation Be Fair? Paradoxes and Possibilities in Fair Trade Darjeeling Tea Certification. *Anthropology of Work Review*, 29(1), pp. 1–9.

Bhullar, L. (2008). The Indian Forest Rights Act 2006: A Critical Appraisal. *Law, Environment and Development Journal*, 4, p. 20.

Borrini, G., Kothari, A. and Oviedo, G. (2004). *Indigenous and Local Communities and Protected Areas: Towards Equity and Enhanced Conservation: Guidance on*

Policy and Practice for Co-Managed Protected Areas and Community Conserved Areas. Kent: IUCN, p. 11.
Brockington, D. (2004). Community Conservation, Inequality and Injustice: Myths of Power in Protected Area Management. *Conservation and Society*, 2(2), p. 411.
Costanza, R., d'Arge, R., De Groot, R., Farber, S., Grasso, M., Hannon, B., Limburg, K., Naeem, S., O'neill, R.V., Paruelo, J. and Raskin, R.G. (1997). The Value of the World's Ecosystem Services and Natural Capital. *Nature*, 387(6630), p. 253.
Daily, G.C. (1997). *Nature's Services: Societal Dependence on Natural Ecosystems* (Vol. 19971). Washington, DC: Island Press.
de Groot, R., Wilson, M. and Boumans, R. (2002). A Typology for the Classification, Description and Valuation of Ecosystem Functions, Goods and Services. *Ecological Economics*, 41(3), pp. 393–408.
Gamble, J.S. (1896). *List of the Trees, Shrubs and Large Climbers Found in the Darjeeling District, Bengal*. Darjeeling: Presidency Jail Press.
Hazra, A.K. (2002). *History of Conflict Over Forests in India: A Market Based Resolution*. New Delhi, India: Julian L. Simon Centre for Policy Research, Liberty Institute.
Holland, B. (2008a). Ecology and the Limits of Justice: Establishing Capability Ceilings in Nussbaum's Capabilities Approach. *Journal of Human Development*, 9(3), pp. 401–425.
Holland, B. (2008b). Justice and the Environment in Nussbaum's "Capabilities Approach": Why Sustainable Ecological Capacity Is a Meta-Capability. *Political Research Quarterly*, 61(2), pp. 319–332.
Holland, B. (2014). *Allocating the Earth: A Distributional Approach to Protecting Capabilities in Environmental Law and Policy*. Oxford: Oxford University Press.
Huntsinger, L. and Oviedo, J. (2014). Ecosystem Services are Social – Ecological Services in a Traditional Pastoral System: The Case of California's Mediterranean Rangelands. *Ecology and Society*, 19(1).
Karanth, K.K. and Nepal, S.K. (2012). Local Residents Perception of Benefits and Losses from Protected Areas in India and Nepal. *Environmental Management*, 49(2), pp. 372–386.
Kothari, A. (2013). Development and Ecological Sustainability in India: Possibilities for the Post-2015 Framework. *Economic and Political Weekly*, pp. 144–154.
Kovac, L. (2012). The Biology of Happiness. *Science and Society*, 13(4), pp. 297–301.
Kulkarni, S. (1987). Forest Legislation and Tribals: Comments on Forest Policy Resolution. *Economic and Political Weekly*, pp. 2143–2148.
Lama, B.B. (2008). *The Story of Darjeeling*. Darjeeling: NYL Publications, p. 167.
Lele, S. and Menon, A. (2014). *Democratizing Forest Governance in India*. New Delhi: Oxford University Press.
Lele, S., Springate-Baginski, O., Lakerveld, R.P., Deb, D. and Dash, P. (2013). Ecosystem Services: Origins, Contributions, Pitfalls, and Alternatives. *Conservation and Society*, 11(4), pp. 343–358.
Lynch, O. and Talbott, K. (1995). *Balancing Acts: Community-based Forest Management and National Law in Asia and the Pacific*. Baltimore, MD: World Resources Institute.
Menon, A., Hinnewinkel, C., Garcia, C., Guillerme, S., Rai, N. and Krishnan, S. (2009). Competing Visions: Domestic Forests, Politics and Forest Policy in the Central Western Ghats of South India. *Small-Scale Forestry*, 8(4), p. 515.
Millennium Ecosystem Assessment. (2005). Ecosystems and Human Well-Being. Washington DC. Island Press.

Mishra, C. (1997). Livestock Depredation by Large Carnivores in the Indian Trans-Himalaya: Conflict Perceptions and Conservation Prospects. *Environmental Conservation*, 24(4), pp. 338–343.

Narain, D. (2015). *History of Forest Governance in India | Divya Narain*. [online] Divyanarain.wordpress.com. Available at: https://divyanarain.wordpress.com/tag/history-of-forest-governance-in-india/ Last accessed on 11 June 2019

Nussbaum, M.C. (1997). Capabilities and Human Rights. *Fordham Law Review*, 66, p. 273.

Nussbaum, M.C. (2011). *Creating Capabilities*. Cambridge, MA: Harvard University Press.

Ogra, M. and Badola, R. (2008). Compensating Human – Wildlife Conflict in Protected Area Communities: Ground-Level Perspectives from Uttarakhand, India. *Human Ecology*, 36(5), p. 717.

Palmer, C. and Engel, S. (2007). For Better or for Worse? Local Impacts of the Decentralization of Indonesia's Forest Sector. *World Development*, 35(12), pp. 2131–2149.

Pelletier, J., Gélinas, N. and Skutsch, M. (2016). The Place of Community Forest Management in the REDD+ Landscape. *Forests*, 7(8), p. 170.

Rai, Y., Chettri, N. and Sharma, E. (2002). Fuel Wood Value Index of Woody Tree Species from Mamlay Watershed in South Sikkim, India. *Forests, Trees and Livelihoods*, 12(3), pp. 209–219.

Rangarajan, M. and Shahabuddin, G. (2006). Displacement and Relocation from Protected Areas: Towards a Biological and Historical Synthesis. *Conservation and Society*, 4(3), p. 359.

Ransome, B. (2010). Sen and Aristotle on Well-being. *Australian Journal of Social Issues*, 45(1), pp. 41–52.

Saberwal, V.K. (1999). *Pastoral Politics: Shepherds, Bureaucrats, and Conservation in the Western Himalaya*. New Delhi: Oxford University Press.

Sandhu, H. and Sandhu, S. (2014). Linking Ecosystem Services with the Constituents of Human Well-Being for Poverty Alleviation in Eastern Himalayas. *Ecological Economics*, 107, pp. 65–75.

Sarin, M. (1998). Community Forest Management: Whose participation . . .? In: I. Gujit and M. Shah, eds., *The Myth of Community: Gender Issues in Participatory Development*. London: Intermediate Technology Publications, p. 11.

Schreckenberg, K. and Luttrell, C. (2009). Participatory Forest Management: A Route To Poverty Reduction? *International Forestry Review*, 11(2), pp. 221–238.

Sen, A. (1999). *Commodities and Capabilities*. Oxford University Press Catalogue.

Sen, A. (2009). *The Idea of Justice*. London: Penguin.

Sharma, C.K. and Sarma, I. (2014). Issues of Conservation and Livelihood in a Forest Village of Assam. *International Journal of Rural Management*, 10(1), pp. 47–68.

Torri, M.C. (2011). Conservation, Relocation and the Social Consequences of Conservation Policies in Protected Areas: Case Study of the Sariska Tiger Reserve, India. *Conservation and Society*, 9(1), pp. 54–64.

Watson, N.L. (2007). Implications of Land Rights Reform for Indigenous Health. *Medical Journal of Australia*, 186(10), pp. 534–536.

5 Our 'Other Species' capabilities and capabilities of other species

Siddhartha Krishnan and Annesha Chowdhury

Introduction

People's exposure to circumstances beyond their control has adverse effects. For human development philosophers, this leads to habituation to negative living environments. People begin to accept their conditions (Sen, 1984, 1999; Nussbaum, 2000). Theirs is an inability to separate habituation or the 'cheerful endurance' of one's circumstance with real happiness (Sen, 1984, p. 309, 1999; Nussbaum, 2000). 'The most blatant forms of inequalities and exploitations survive' by 'making allies out of the deprived and the exploited. The underdog learns to bear the burden so well that he or she overlooks the burden itself' (Sen, 1984, p. 309). Acceptance replaces discontent and 'hopeless rebellion by conformist quiet' (Sen, 1984, p. 309). Cheerful endurance replaces 'suffering and anger' (Sen, 1984, p. 309). 'As people learn to survive to adjust to the existing horrors by sheer necessity of uneventful survival, the horrors look less terrible in the metric of utilities' (Sen, 1984, p. 309). Are you then satisfied, less satisfied or least satisfied with compensation for losses caused by wildlife?

Capabilities is about opportunities and freedoms to pursue valuable things. The key here is 'value'. And attitudes are not values. If a positive relationship with wildlife is a capability, then negative attitudes deter robust assessment. A negative attitude towards a leopard may arise in the immediacy of a raid or in stressful circumstance of exclusionary protected area management. In understanding wildlife depredation of well-being resources, and wildlife as a well-being resource, values are important. The sociologists Steven Hitlin and Jane Allyn Piliavin critique conflation of value and attitude (2004). Attitudes are 'favourable or unfavourable evaluations of an object' (Hitlin and Piliavin, 2004, p. 361). Say a raided crop or crop raider. 'Values focus on ideals; attitudes apply "more to concrete social objects"' (Hitlin and Piliavin, 2004, p. 361). In an ethical context, ideals are values that one pursues. Attitudes pertain to more material pursuits. They cite two classic definitions of values. Kluckhohn's behavioural definition and Rokeach's general definition. For the former, value is an individual or group 'conception, explicit or implicit' 'of the desirable' that influences action (Kluckhohn, 1951, p. 395). Values for the latter are enduring beliefs over preferred personal or social conduct (Rokeach, 1968). Values were action for Kluckhohn; values lent meaning to action for Rokeach. Meaningful thought and action are more dependable well-being information in human-wildlife conflict contexts. While attitudes may reveal values, the latter are more durable. Values stand more implicated in personhood and less in behaviour. Aristotelian quality-of-life evaluations have personhood or quality of individual basis.

This research on how values form and change takes us to the chapter's ancillary question. Are positive and negative opinions and perceptions on wildlife adaptive preferences? These opinions and perceptions help us assess 'other species' capabilities (Nussbaum, 2011, p. 34). But what is the conservation context to critically locate other species capabilities? This then is the main question about people's capabilities to show wildlife concern and attachment. Nussbaum's 'other species' include animals, plants and nature. We engage with ecological capabilities and contexts in previous chapters. In this chapter, we discuss other species' capabilities in the context of wildlife conservation, not animal welfare. In fact, we examine debates on the differences between urban animal welfare and rural wildlife conservation. In the temperate north, where people do not normally live in or next to Protected Areas, there are low risks and costs of conflating welfare and conservation. But in tropics like India where millions live in or near forests, confusing animal welfare with wildlife conservation renders human lives precarious. There are dog people, cat people and tiger people. If the former two and the latter confusedly form an ideological coalition, *Adivasis* become an isolated collateral.

This is a good time to signpost the chapter. In the following section, we set the 'conservation' context for Nussbaum's other species capability – the context being human-wildlife conflicts. We discuss a few reviews of the field and pertinent Darjeeling studies. We examine and critique the reviews for their inadequate use of 'attitudes', 'perception' and tolerance. We contend that such 'fleeting' subjective research is blind to adaptive preference. We then discuss Nussbaum's other species capabilities, especially of wildlife. We then invoke an environmental sociology of representation that provides a framework to critique wilderness. We then differentiate wildlife conservation from urban animal welfare. The following section examines empirical conflict realities around the two Protected Areas. Here we compare conflict, capability and adaptive tendency in Lalung and Rammam villages. We then conclude with a discussion on other species and material and political capabilities.

Human-wildlife conflicts

Human and wildlife conflicts are a stressful conservation context to evaluate 'other species' capabilities. Paraphrasing, in Singalila and Senchal, can people live with concern for and in relation to leopards and wild boars? Conflicts (some contest this term for agency afforded to animals) with wildlife are prosaic around PAs. Wildlife and forest laws have outlawed 'effective' mitigation measures like hunting or changing habitat. Social scientists and conservation biologists have filled this conflict space with publications. Conflict studies by species, space, perception, loss and mitigation abound (Rai et al., 2014; Chakraborty et al., 2012; Parvaiz, 2017). The few studies of conflicts in Darjeeling and reviews of the field in general suffice to set context.

First, the reviews. Usually comparative, they engage with people's 'attitudes' towards depredators. Treves et al. (2006) for instance in their study of wolf conflicts in Bolivia, Wisconsin and Uganda, encourage the use of social sciences especially in studying people's risk perceptions. Conflict co-management that makes for viable politics need measure perceptions besides participatory planning. The authors warn against 'skepticism' of perceptions and attitude not because they could be adaptive, but because they reveal concern over certain species. Catastrophic events may shape these perceptions (Treves et al., 2006, p. 387). Here, a negative perception of wildlife may appear 'non-adaptive'. People may appear objective and truthful about their experiences and thoughts. But stressful conservation and conflict circumstances may even stifle positive perceptions of wildlife. Such 'good' thoughts about wildlife may become causality to oppressive conservation circumstances. A forest dweller's empathy for boars and bears may recede because forest laws delay or deny crop raid compensation.

Treves et al.'s word of caution to ecological sciences about scepticism over perceptions is significant from a capabilities perspective (2006). More useful, they argue, is seeking explanations for concerns over conflict species. So is their emphasis on perception's role in gauging 'tolerance' for conflicts. Perceptions shape tolerance for 'interventions', be they mitigative, including tolerance increasing (Treves et al., 2006, p. 388). Such caution and emphasis is important in a field dominated by natural sciences. But scepticism and tolerance from a capabilities perspective pushes social science possibilities. For scepticism here is also about ontology, besides epistemology. Perceptions are opinion data on wildlife. But also subjective opinion that needs objective assessment for their 'forced' formation. Tolerance, then, is not patience or impatience. It is favourable or unfavourable animal appraisals under hostile conservation circumstances.

A review of human and felid conflicts emphasizes the significance of perceptions and attitudes. Also, their links with tolerance. For conservation biologists Chloe Inskip and Alexandra Zimmermann, attitudes and perceptions are important socio-cultural factors. Attitudes and perceptions determine tolerance and govern levels of human response to felids. They may distort conflict scale. People's retribution is disproportionate to the scale of the wildlife problem (Inskip and Zimmermann, 2009). But tolerance for a leopard, and the attitude scale that connotes it, may conceal – or reveal – adaptive preference. People's anger and love for leopards could well be mental states wrought by restrictive conservation laws.

Latent scope for objective appraisals of subjective data remains, appraisals that infer adaptive preference and establish context for it. Such scope emerges in the reframing of human-wildlife conflicts as people-people conflicts. Conservation biologists Stephen Redpath et al. reviewed 100 human-wildlife conflict papers since 2010. Language 'matters' they infer (Redpath et al., 2014). The human-wildlife conflicts examined were between conservation and livelihoods defenders and also conservationists versus common folk. Technical solutions follow the human-wildlife conflict trope – fencing and early warning systems, for instance. But these conflicts are between people. Technical solutions are 'unlikely to focus on the underlying problem'. Because a technical solution is effective, conservation and livelihoods conflict need not recede. This conflict reframing offers latent or untapped potential for exploring adaptive preference. It is not manifest yet. For instance in acknowledging that problem framing has repercussions, the conservation biologists write 'presenting wildlife as antagonistic may alter the way people perceive those species' (Redpath et al., 2014, p. 2). This is a step away from realizing that people's species perception is altered by antagonistic circumstances wrought by conservation policy. Both cheerful and harmful dispositions to wildlife could be 'normalizing' or adaptive behaviour.

The few studies from Darjeeling focus on material deprivations in indifferent conservation contexts and also on absence, soundness or success of technical solutions. Roshan Rai et al. (2012) report on conflicts in four Protected Areas in Darjeeling and Sikkim. In Darjeeling, Singalila National Park forms the study area. At the receiving end are subsistence farmers with small rain-fed acreages. They face depredation by non-charismatic species including small mammals and herbivores. These include barking deer, porcupines and yellow-throated martens. In Singalila's Samanden village, wild boars are the main crop raiders. In 2011–2012, farmers lost 45.5 percent and 81.7 percent of their potato and maize crops. Farmers reason that boars are short of food in the forests. Sikkim has a compensation distribution scheme that is callous in costing and outreach. But in Darjeeling's Samanden, there is no such scheme (Chakraborty et al., 2012).

Reframing human-wildlife conflicts as human-human conflicts is difficult as Redpath and colleagues argue. This is particularly so when charismatic animals are depredators. Though counter-intuitive, the less charismatic is the animal, the tougher is the reframing of conflict as a human-human one. Why will one configure larger conflict of human interests when a wild boar is the depredator? This is why Roshan Rai and colleagues may not sense adaptive preferences of their subjects. But they get there. They do knock on doors that shut out choices and opportunities for forest-edge farmers. Communities bear the brunt of conservation discourse and efforts, they state. Their 'voices and difficulties are going unheard and unattended'. Bearing the brunt is about suffering the worst part of something and putting up with it.

As they conclude, conflicts impact marginal communities with 'no reported instances of retaliation' (Chakraborty et al., 2012, p. 7). Also, with traditional livelihoods uprooted, migration is an adaptive measure.

But absence of retaliatory intent, let alone action, need not mean empathy or tolerance. Retaliation also does not mean total animus or intolerance. Further, not migrating need not rule out any form of adaptation. The farmer may subsist hopeless and powerless at forest edges. Physical migration is an extreme decision. One can migrate mentally from aspiratory to acquire mind-sets.

In a later paper, Roshan Rai and colleagues emphasize presenting and leveraging stories – the story of conflicts involving non-charismatic species, not the common 'big' problem animals. This and the assumption that damage by small mammals is insignificant ensured such stories did not cross 'mid-management' levels in the Forest Department. Officials questioned these stories and de-prioritized them. These stories backed by hard scientific data collected by various state and non-state actors increased visibility of the conflicts. The mitigative and investigative techniques the authors recommend are amenable to a certain rhetorical flourish. To keep crop raiders away, the authors note the efficacy of bio or living fences made of plants. Such interventions are 'interdisciplinary requiring diverse knowledge and skill sets'. Human-wildlife conflict management in Darjeeling and Sikkim call for interdisciplinary efforts (Rai et al., 2014, p. 80).

In the spirit of interdisciplinarity one must look beyond utility and engage with mentality. If bio-fences are a mitigation technique, so are mental-fences. People can mitigate in their heads. They can endure and accept crop raids and policies that enable them. They can fence out aspiration and fence themselves into resignation. Roshan Rai et al. do not discuss any perception and attitude data for us to consider. But if they did or do discuss people's opinions, adaptive preferences could be unveiled. While bio-fencing is of livelihood significance, a focus on technical solutions is inadequate. As Redpath et al argue, if we view the conflict as humans versus wildlife, the focus is on technology rather than 'underlying conflict'. Technical solutions reduce impacts but conservation versus livelihood conflicts are not addressed (Redpath et al., 2014, p. 2). In Darjeeling and Sikkim, the wildlife may be small, but the bigger conflict remains.

When Redpath and colleagues reviewed human and wildlife studies, other underlying conflicts and issues were evident. Besides conservation and livelihood interests, the issues were about 'animal welfare' and 'human well-being'. Animal welfare emphasized moral responsibility towards conflict species, especially in urban landscapes. As for human well-being, the emphasis was also on people's religious connections with species. This is a good point to discuss Nussbaum's other species capability. We must locate this capability in conservation contexts discussed previously and also in critiques of bio-centric conservation and conflation of wildlife conservation with animal welfare.

Nussbaum's other species capability is the ability to 'live with concern for and in relation with animals, plants, and the world of nature' (Nussbaum, 2011, p. 34). But our species concern here is animals, especially non-charismatic wildlife!

The capabilities approach is also referred to as the human development approach. But Nussbaum prefers capabilities over human development for an approach nomenclature. Non-human animals play a role in this preference. Nussbaum has concern over non-human animal capabilities besides those of humans. For her non-human animals have entitlements. Like humans have under the Humans Rights Declaration. For Nussbaum, the state must enable its citizens to have a minimum ability for animal concern and relationship. Among other capabilities this is necessary for dignified living or flourishing. Would the fact that she places animals along a species spectrum including plants and nature mean that by animals she means wildlife? And not pets like dogs and cats? Or that by animals she means both wild and domesticated? This is important to clarify given conflation of wildlife conservation and animal rights in India.

This conflation has crept into conservation debates around Protected Area policies. These promote and pursue anthro-sanitized 'wild' spaces for big mammals. Entitling wildlife in India has resulted in forest dweller disentitlement. The Forest Rights Act's entitlement of *Adivasis* and other forest dwellers has potential to create a cultural and political environment for species concern. The Act is a political environment – a political environment and other species capability – for converting political resources into valuable functioning or being and doing concern for wildlife. But bio-centric conservationists and the capitalist state seek to dis- or non-entitle; the former demanding exclusive protection space and the latter transforming space by changing land use. Re-notifying a forest as a critical wildlife habitat or de-notifying a critical wildlife habitat are options that have been exercised. Such exclusion or transformation require denying forest rights or depriving granted rights through creative legislation. For instance recent proposed amendments to the Indian Forest Act seek to provide for commuting rights granted under the FRA.

Nussbaum's other species capability needs contextualization not just because it is abstract and needs to be set in 'the narrative context of human lives' (Nussbaum, 2011, p. xi). But also because narratives on forest contexts record the fate of human lives. These critical narratives note the livelihood consequences of people's non- and dis-entitlement as a wildlife protection measure. In capabilities, implicated are capability deprivations – specifically the material and political control capability deprivations wrought by bio-centric species ideologies. The threats *Adivasis* and other forest people face from wildlife are not just raids, but resettlement. Resettlement is not exactly rehabilitation. The latter, involving creation or substitution of meaningful livelihood opportunities, has a bad record in India.

Overall this conservation context renders challenging the identification or recommendation of a threshold for the capability of showing concern for wildlife or living in relation with them. Love or loathing for leopards may well turn out to be adaptive preference data. As stated earlier, attitude and perception surveys are insufficient but useful methods to assess adaptation to circumstances. At this surveyed level, *Adivasis* and other indigenes may express satisfaction or dissatisfaction over wild animal raids. Utilitarian surveys by conservation biologists invoke raid events and suppose that the extent of economic loss will influence the scale of pleasure or pain that 'respondents' feel. Income deprivation does lead to capability deprivations like illiteracy and ill health as Sen agrees (2000). But income need not be the only informational basis for good or bad feelings for animals. Political deprivations, or from Nussbaum's list, control over one's material and political environment capability, are implicated in animal feelings. Land and participation entitlements are strong foundations for feelings. There is the methodological need to cease remaining feeling surveyors and become more sound scholars of depriving contexts.

For feelings are fleeting psychological states. They are reflexive automatic responses to animal atrocities. Not reflective thinking. For Sen, 'reflective valuation' is ignored under welfarism. Welfarism chooses subjective utility as information and ignores valuation. By this he means both deeper evaluation by self, and objective evaluation by the scholar or 'expert'. Subjective welfare then is vulnerable to the adaptive preference of people vulnerable to wildlife.

Nussbaum's critique of utilitarianism invokes another state from which powerless poor people suffer. It complements vulnerability in reducing forest folk to pitiable proportions. Engaging with Sen's critiques, she notes that preferences and satisfactions are socially malleable. So forest folk are exposed to wildlife and any 'tolerant' or pro-animal position they convey are shaped by punitive and exclusive forest policy. And as we suspect, any anti-animal position is also a malleable one, shaped by anti-establishment sentiment. 'When society has put some things out of reach for some people, they typically learn not to want those things', writes Nussbaum. These 'things' can be *concern for wildlife*, or *concern over wildlife*. More generally, the 'things' beyond reach for Indian forest dwellers have been development things – education, dispensary, electricity and roads (Nussbaum, 2011, p. 54). If any study suggests or holds that forest dwellers do not want any of these 'modern' things, then chances are that they learnt to not want to. Behind any ostensible anti-modern posturing could well lie a tradition of denial.

Before we compare the vulnerability and malleability of Lalung and Rammam folk to wildlife raids and wildlife protection policy, let us review Nussbaum's other species capability. And her entitling of animals through an 'expanded notion of dignity' (Nussbaum, 2011, p. 161). Nussbaum also critiques 'wilderness' from an animal capability point of view. Dignity and wilderness also figure in critiques of the Protected Area philosophy. We review some pertinent critiques. These provide a substantive political context in which one can locate Nussbaum's animal philosophy.

Dignified humans and animals

Dignity, as discussed in the 'Justice and the Jungle' chapter, plays a foundational role in Nussbaum's capability list. She invokes dignity when discussing a human life worthy of it, or the dignity appropriate to a non-human species. Showing concern over and having a relationship with animals is a capability. Its minimal achievement or possession is a requirement for a dignified human life. Proponents of the capabilities must take a 'fundamental' position: 'Whose capabilities count?' Proponents will hold that capabilities of all humans equally count. But there are five different positions beyond this. We paraphrase Nussbaum's positioning using Himalayan human and non-human and mammal examples.

1. *Sherpa* capabilities are ends in themselves, though red panda capabilities may be instrumentally valuable in promoting the former's capabilities.
2. Yak roles in relationships with *Lepchas* are not just means but have intrinsic value to the relationship.
3. Capabilities of all Himalayans, human and non-human, as ends in themselves and requiring threshold attainment.
4. Capabilities of rhododendrons and yellow-throated martens matter, but as individual entities and not as ecosystem constituents.
5. The alpine and sub-alpine ecosystem capabilities count as ends in themselves.

In her *Women and Human Development* (2000), Nussbaum took position 2, that human capabilities include relations with species and nature. Other species are not instrumental to this relationship but are constitutive of it. In her *Frontiers of Justice* (2006), she defended position 3 on a social justice count. Animals are sensitive to pain. Like humans, they can feel pain. They strive to live like humans and can find their pursuits 'thwarted'. In recruiting animals into the social justice ambit, Nussbaum makes a novel distinction between animals, trees and ecosystems. The latter two do not 'suffer injustice'. They can be damaged and there are moral and non-moral reasons for not damaging them. But they do not experience pain. Ecosystems are 'not a center of experience' and 'do not have a life-project or striving'. So, for Nussbaum it would 'seem odd' to suggest that, for example, Himalayan oak forests 'suffer injustice'. Non-moral caring of the 'intellectual, scientific and economic' kind remain for damaged trees or ecosystems. These 'intuitions are difficult to articulate, and only time and future debates will more clearly map this complicated terrain' for Nussbaum (Nussbaum, 2011, p. 159). In India, recent legal adjudication of rivers as persons with rights is one such debate. As is the debate following the discovery that trees have a network of roots through which they communicate. But this is a matter of another chapter, another book.

Suffice that the court's was an expansive notion of justice, like Nussbaum's 'expanded notion of dignity' (Nussbaum, 2011, p. 161). Capability talk must be inclusive and not confined to human dignity alone. The dignity of diverse sentient creatures matters too. The capability approach considers every animal type has dignity by itself. Respecting that dignity does not depend on any human obligation. For instance, Kantian approaches, like Christine Korsgaards', treating animals well is a duty deriving 'from duties to support our own human animality' (Nussbaum, 2011, p. 161).

As humans, our duties are towards fellow individuals. For animals, the 'species' provisions characteristic life form to be promoted. Humans should promote opportunities according to each species' form of life. By this measure, urban pet dogs and wild forest elephants exhibit typical life forms. Humans must take into account their distinct or different forms of lives in promoting opportunities. For Nussbaum, we can consider choice if animals exhibit capacity for it. Otherwise, their functioning can be our focus. A 'kind of sensitive paternalism' is appropriate in the animal case (Nussbaum, 2011, p. 162).

Paternalism is an 'authoritarian' limiting of agents' freedoms in their own interest. Environmental sociologists and historians have termed authoritarian efforts to protect wildlife in de-populated national parks. There are no pure wild spaces in the tropics for wildlife to be exclusively protected in goes the powerful political ecological critique. We will discuss these human rights positions soon. But Nussbaum critiques 'wilderness' from an animal welfare or rights perspective. She rejects as naive the 'romantic naturalism' that animals be left alone if they are in the wild. Human action is pervasive in every habitat. Pretending that African elephants 'are in the wild' ignores the fact of 'human plans' encroaching habitat. Only wise intervention can ensure elephants a 'decent shot at living'. Nussbaum prioritizes habitat protection, but holds that well-protected habitats cannot 'support expanding populations'. Animal contraception was a possible intervention option (Nussbaum, 2011, p. 161).

Wilderness and its representation

Nussbaum's critique of wilderness is an aggregating one. Human actions and plans in the tropics that a capabilities scheme need consider are those of colonial and post-colonial conservation policies. The introduction chapter discussed the social and ecological costs of colonial forestry and post-independence Protected Area laws. In the tropical south and its temperate regions, the actions and plans of colonial and contemporary conservation actors caused massive capabilities outages among local resource cultures. All in the guise of, or some genuine love for, wildlife. Peasants and pastoralists have suffered the label of 'encroachers', a materially and politically disentitling legalese. Romantic naturalism has also been a human rights–based critique, just as it is here an animal rights–based critique. The wilderness ideal originating in the west has ostracized local people from 'pristine' rainforests and savanna or has orientally imagined them as part of wilderness.

In fact, just like letting *wild animals be* is naive from not just an animal capabilities perspective, but also from a human capabilities perspective. There is no non-human wilderness for animals to be let wild. Elite human action in the guise of wildlife preservation has been pervasive in every third world habitat. Conservation deprivations abound. Farmers, grazers and foragers have either been displaced externally or internally; the former entailing resettlement and no substantive rehabilitation and the latter entailing disentitled and precarious persistence in *resource-rich* habitats. Here the 'protected' environment itself becomes a hurdle depriving opportunity to convert resources into capabilities.

The capabilities approach may consider animals as agents and 'not as receptacles of pleasure or pain' (Nussbaum, 2011, p. 160). But the elephant or tiger are incapable of unambiguously communicating to humans their habitat needs and interests. They need to be 'represented', for instance as Nussbaum suggests, through 'paternalism'. Romantic naturalism is also a representation of the wild. We are discussing human and animal well-being relationally. Even the notion of ensuring animal capabilities for intrinsic own-sakes is an essentialist human position. And animal caring is supposedly representative of human goodness. So, representation has this mutually constitutive nature. The represented and representative as a relational interface play important environmental and well-being roles. All the same, representation poses well-being risks. The power that rich races and classes have in representing geography as wild, or local communities as cultural threats to the wild, can render precarious these communities. Human-rightist representations of animals as pestilent and depredatory also has implications for animal well-being. The environmental sociology of representation is an appropriate political field to discuss human and animal entitlements.

Scandanavian sociologists Magnus Boström and Ylva Uggla draw upon H. F. Pitkin's classic 1967 work on representation. They make a case for its salience for the environment. Pitkin's ideas of 'standing for' and 'acting for' are two facets of representational practice. The image of the polar bears on melting ice *stood for* – represented – climate change. Al Gore, the director of the documentary *An Inconvenient Truth* where the image was used in animated form, *acted for* the environment. Pitkin emphasizes representation as acting for others and in the interest of someone or something else. This includes 'acting in someone's interest even if his/her wishes are not or cannot be articulated', wishes that are potentially relevant (Boström and Uggla, 2016, p. 356). Also, acting to represent abstract ideas like justice. Animals cannot articulate their wishes, and their wishes are relevant for justice. Philosophers like Nussbaum articulate animal interests from a justice perspective.

A justice and dignity-based articulation is necessary and novel in the polemical field that is animal welfare. And so, animals are not really a 'missing environmental constituency' in the field of representation. The Scandinavians cite the political scientist Andrew Dobson's (1996) listing of such constituencies, which includes future citizens, and along with animals, whole natural environments. Much speaking up or standing in for happens in the arena of environmental problems. But this abundance has not been 'sufficiently understood as representation' (Boström and Uggla, 2016, p. 356). There is also a representation theme of political sociological interest. Contemporary representative democracy is insufficient to deal with environmental problems. They cross borders and have long-term implications. Not all affected are able to participate in 'relevant political forums'. They need to be represented. A solution offered by Dobson is identifying a 'sustainability lobby' that represents 'foreign affected citizens', 'future citizens and animals' (Boström and Uggla, 2016, p. 358).

The problem of representative democracy is also about environmental justice. But not substantively. Toxic wastes can be imported to the third world with health risks. Carbon emissions in the north can have climate consequences for a tropical island. Suffering waste-sorters and fishermen need representation in national political institutions and international forums. Perhaps sustainability lobbies can represent people disproportionately facing health and climate risks. But we must have better representations of environmental problems and injustice. Efforts by a northern lobby to create exclusive Protected Areas in the south is an environmentally unjust problem. International lobbying for militarization of elephant conservation in Africa or rhino conservation in India is also a cross-border environmental problem. Its long-term consequences for local cultures can only be displacement and deprivation.

Sociologists Boström and Uggla realize this power play in representation. For them, representation can be opaque and unreflective, and 'produce and reproduce power relations'. Environmental sociology need focus on the practices of environmental representatives (Boström and Uggla, 2016, p. 359). Landscapes can be represented wrongly and the corrective is attention to local representations. Guinea's is a classic case of such misreading and misrepresentation. Geographers James Fairhead and Melissa Leach critique scientist and policy maker worry over 'savannisation' in Guinea. A discourse since French colonization of Guinea that claimed humid forests were original habitat now reduced to riverine and swampy patches. Shifting cultivation and firing were blamed. That forest is 'nature' and use degrades upends in the experience of Guinea's Kuranko people. 'Untouched land tends to be savanna, and that where one cultivates, trees multiply'. Farmer representations of the landscape and its transformation contrasted scientific and policy knowledge, the latter linked intimately to intellectual and financial-institutional structures. The degradation discourse was Foucauldian in its powerful exclusion of alternative landscape representations. But there was no homogenous local perspective either. A 'plethora' of landscape representations existed. The landscape was at once a material and discursive outcome (Fairhead and Leach, 2000, p. 14).

Wilderness discourse has capability consequences. The Indian state created 105 national parks and 543 wildlife sanctuaries. Political ecologists

Daniel Brockington and James Igoe cite Ashish Kothari's (2004) estimate that approximately 4 million people were vulnerable to eviction due to stricter PA laws (2006). The category Wildlife Sanctuaries includes 50 Tiger Reserves under the aegeis of a 'Project Tiger' policy. Project Tiger and Project Elephant were state genuflections to representations by first world experts. Conservation biologists, the tiger's and elephant's expert representatives, have not been 'sensitive paternalists'. The historical sociologist Ramachandra Guha called them 'authoritarian biologists' (Guha, 1997, p. 14). These scientists measure forest dweller patience with wildlife today. But for Guha impatience with 'humans is especially marked among conservation biologists for whom farmers and forest-dwellers represent a messy obstacle to the unimpeded progress of scientific research'. It was the white biologist's burden to know abandoning home, field and forest was in the 'natives' true interest'. The new rulers were 'not the animals they once co-existed with, but the biologists, park managers and wildlifers'. They determine how territory is to be managed (Guha, 1997, pp. 15–16).

Material mismanagement of territory is one thing but metaphysical mistaking of territory is quite another. And metaphysical mistaking also has material implications. Animal welfarism has encroached on wildlife conservation territory in the guise of 'compassionate conservation'. It is 'flawed' and dangerous for people, wildlife and ecosystems, argue environmental historian Meera Oommen and her political ecologist colleagues. Compassionate conservation for Arian Wallach and colleagues (2018) is safeguarding biodiversity 'while retaining a commitment to treating individuals with respect and concern for their well-being'. They propose a 'universal moral based on sentience'. Key tenets include 'do no harm' and 'individuals matter' (Oommen et al., 2019, pp. 1–2). The capacity for pain and pleasure, and the capacity for active striving, brings animals under a social justice ambit says Nussbaum. Humans and animals can experience pain and harm, and both attempt to live and act. These are unjustly thwartable projects. As a concept, justice is tied to the idea of 'experienced harm and thwarting' (Nussbaum, 2011, p. 158).

But conservation, whether compassionate or scientific, harms and thwarts humans. Oommen et al. weigh the practical costs of 'overplaying moral salience of sentience and sapience'. They cite the political ecologist Roderick Neumann's work on violence in new moral geographies in Africa. Humans suspected of poaching are executed without due process. Humans without control over material and political environment capabilities. They are often products of 'difficult social, economic, and political circumstances well beyond their control' (Oommen et al., 2019, p. 2). Wallach et al. (2008) use India as a model for compassionate conservation. India's is a successful combination of principles of compassion and conservation. They highlight its vegetarian tradition. This representation is fallacious in its assertions about meat eating. The conservation model is 'compassionate and successful only if enduring and widespread brutality, impoverishment, coercion, and exclusion of marginalized communities by elites are ignored'. Combining wildlife preservation and compassionate animal rights is problematic in the context of human-wildlife conflict. Hundreds of humans die every year along with a few elephants. Yet there is reverence for animals and is part of a 'multifaceted' relationship. These range from 'eliminating problem animals, hunting for meat and sport', and deriving 'religious and spiritual succour and companionship'. Conservation must be responsive to empirical reality, political contestations and real world complexity (Oommen et al., 2019, p. 2).

Nussbaum's compassion for animals is strong, as is her compassion for people. But her social justice pitch for individual animal capabilities is elitist. What else can it be if we weigh Guha's, Neumann's and Oommen et al.'s criticisms of wildlife representations? There is also the general critique of the liberal universalism undergirding her capabilities approach as being as elitist. But in making a universal pitch for human capabilities Nussbaum actually makes an anti-elitist case. Local narratives help communicate 'life conditions' she writes. They help others imagine and acknowledge 'equal humanity' of lives ignored by 'privileged elites'. They also locate capabilities in human narrative contexts and show the difference it makes for what policy makers notice in local lives. Policy interventions will be respectful of real people. It will not reflect 'biases of intellectual elites', Nussbaum states. Yet, wildlife policy whether scientific or compassionate, thwarts life conditions of real people (Nussbaum, 2011, pp. xi, 81).

Let us get to the 'empirical' and complex 'narrative' contexts of Lalung and Rammam (see Figure 5.1). Besides small mammal depredations here, the complexity would be adaptive preferences of people. Their likes and dislikes of wildlife raises suspicion of habituation and adjusting. We are assessing their capability to 'live with concern for and in relation with animals'. We ask about their relationship with the wildlife and seek to understand what concern or contempt for animals is discernible from this relationship. Both concern and contempt for wildlife can involve adaptive preference.

We introduce, in brief, Rammam and Lalung society and history. We then provide simple percentile measures, first, of respondents who experienced crop and livestock raids. We then describe the harsh materialiaties of depredation, including material and political environment capability deprivations. We then provide a percentile measure of those who show concern or contempt of wildlife. We then conclude comparing depredation, deprivation and habituation in Lalung and Rammam.

Lalung and Rammam: depredation and deprivation

In Singalila's Rammam, and Senchal's Lalung, community interactions with wildlife were conflictual. People live in proximity to the two PAs, but do not have security of tenure or participatory forest management rights. This was exposure to adverse circumstances. It can affect their ability to develop

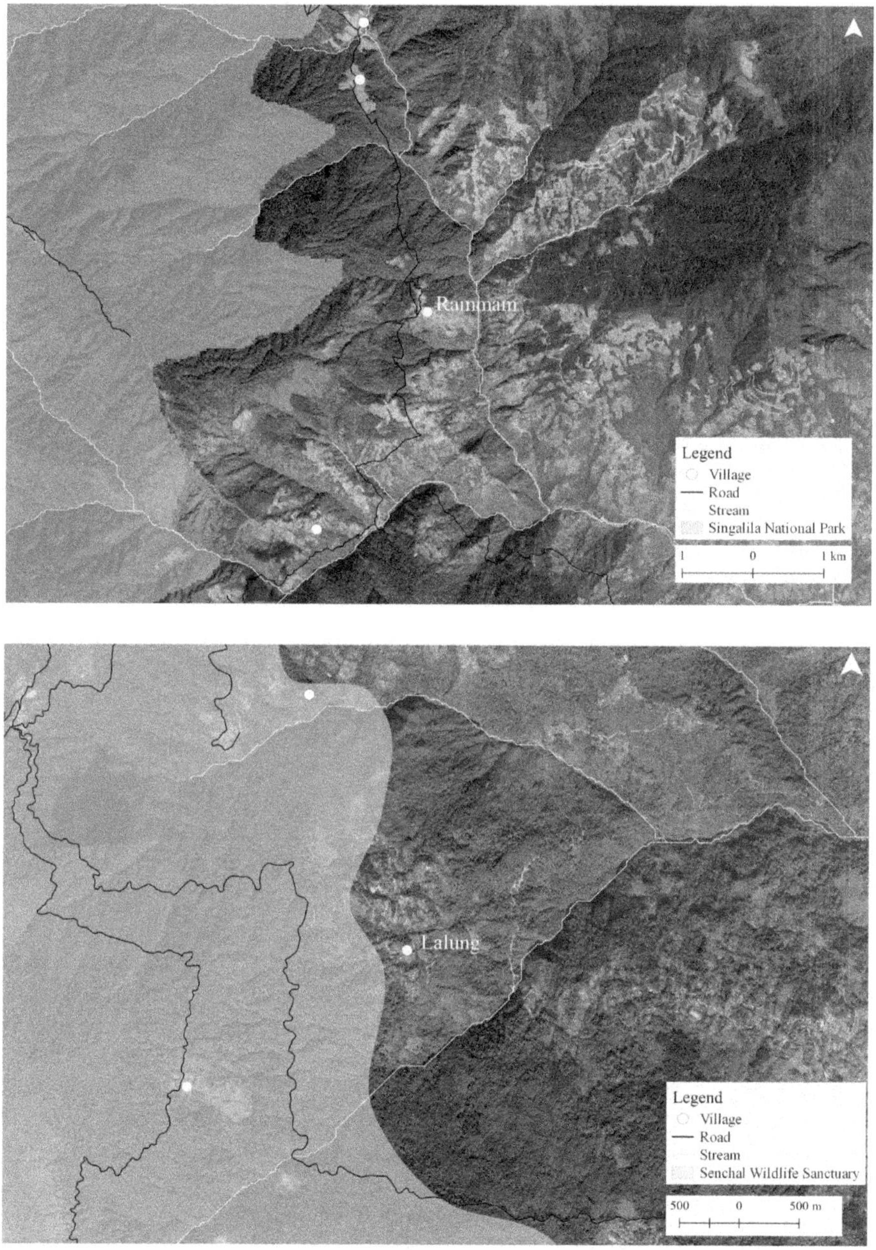

Figure 5.1 Rammam and Lalung: on the fringe

Source: Generated by Abhishek Samrat.

objective opinions of wildlife and nature. We will see. Both villages experienced frequent and intense crop raids and livestock lifting. Crop raids was the bigger stressor. A large proportion of people have stopped or reduced livestock rearing. There is a ban on grazing and restricted entry after Protected Areas notification. Rammam and Lalung are hamlets created by British foresters. They had to fell forests for monocultural plantations in the early 1800s.

Rammam is a forest village in Singalila National Park's North range. It falls under the administrative block of Darjeeling-Sadar sub-division, Darjeeling district. It has 52 Buddhist, Hindu and Christian households. In ethnic terms, *Rais*, *Chettris*, *Tamangs*, *Thammis* and *Subbas* reside here. The Forest Department controls their access to forest goods and services.

Lalung is a plantation village, a settlement established for Cinchona plantations around 1862 (Darjeeling.gov.in, 2019). Lalung is near Mungpoo, a revenue village near Senchal Wildlife Sanctuary. It falls under the Darjeeling sub-division. Like Rammam, its people have mixed religious and ethnic identities. *Rais*, *Tamangs*, *Dukpas* and *Subbas* live in Lalung. Our interactions with Lalung's people revealed a jurisdiction anomaly. Lalung does not fall under the purview of the forest or Cinchona departments. But their access to forest resources is restricted and monitored by the department and the plantation.

Rammam and Lalung communities, then, have lived in settlements for generations. They have no land entitlements. Their main source of income is agricultural. A large proportion of the population have stopped or reduced livestock rearing. This, owing to the ban on grazing and restricted entry into the forests since the forests became notified as 'Protected Areas'.

Approximately 75 percent of respondents from Lalung and 97 percent of respondents from Rammam report crop raids and livestock lifting. Agriculture and cattle rearing had become difficult and frustrating. A small but substantial portion of people (18 percent) do not clearly state crop damage as being a major problem in the Senchal area. A good percentage of farmer-respondents feel that wildlife should be conserved and the forest is important for them despite the crop raids and losses. In Senchal, 43 percent of respondents who face crop raids felt that wildlife must be conserved whereas in Singalila, it is 60 percent.

It is tempting to infer that farmers who say good things about wildlife despite the depredations have an other species capability threshold and that farmers who are disparaging of wildlife as suffering a low capability on this count. But animal concern can be adaptive preference given the conservation circumstances. And so it is tempting to treat the positive animal evaluations of suffering farmers as adaptive preference. And the negative animal evaluations by suffering farmers as authentic and natural. But negative dispositions can also be choiceless and forced. People forced to not like animals because conservation policy does not help mitigate or compensate conflicts. But given the material conditions of depredatory deprivation, it is important to suspect adaptive preference in both good and bad things people think and say about animals.

First, let us look at the material circumstances in Lalung and Rammam. In Lalung, deer and rabbit have compelled people to stop growing peas. It is hard work to grow peas and protect it. Raids make it more labour-intensive. Like peas, cabbage cultivation has also stopped. Bears and boars raid maize. Monkeys raid rice and flour. They also join the maize raid party and are known to attack people. Leopards lift chicken, dogs and goats. Maize cultivation has decreased due to boar raids. There is resignation. One cannot stop animals, farmers say. They say they have no idea how to scare away the wildlife.

Despite a year of hard work, the crop is destroyed and labour goes to waste. People have lost patience and capability to endure and persevere. A Dukpa farmer gives a sense of wildlife thwarting farmer flourishing: 'Main problem is animal entering'. 'Otherwise can do anything'. Fencing has farmer consensus as mitigating technology. But the Forest Department needs to approve and also fund a good-quality fence. Forest and wildlife laws thwart any solution. One senses hopelessness among farmers.

In Rammam fences keep porcupines away. But they do not work for wild boars farmers say. And monkeys are tough to keep away. As per a *Rai* farmer, 'sometimes when we sow potatoes, the wild boars come. When the corn ripens, then bears come. Monkeys come, deers'. Maize is a staple crop, and a *Tamang* farmer notes the effects of depredated yields. Farming has become difficult for villagers, he says, which is why villagers have reduced the scale of farming. They are 'going out to do non-farm activities. Earlier they used to eat, sale and prepare alcohol from the maize that they used to grow. Now they do not even roast maize not that frequently'. There is nothing one can do, a *Rai* farmer says. Another *Rai* farmer says, 'Solution? No there is no solution. If they come then they eat and go'. This easy and destructive will and agency of animals is evident from a *Tamang* farmer, 'They come eat and enter the forest. Nothing to stop them'. Another *Tamang* farmer responds to a question of protecting crops from animals: 'Chase them with fire; beat with tin'. But these are ineffective, he claims. Safeguarding crops is as laborious as sowing and harvesting, it appears.

While cultivation is a valuable activity and constitutive of flourishing itself, raid vigilance is forced functioning. Farmers run, chase, beat tins, set fire, erect fences and lose sleep. The nocturnal vulnerability of Rammam's people is starkly evident in their mention of the 'night'. Farmers lose sleep. Losing sleep watching out for animals, or worrying about them, impacts one's internal, and by extension, one's external capabilities. The nocturnal vulnerability of farmers has had an ironic effect of day-time vigilance being preferable. There is a need to keep tight vigil, a *Rai* farmer observed. 'So during the day we manage to take care but at night there are a lot of raids'. A *Chettri* farmer says, 'In the afternoon we can protect the crops from the monkeys. But it is hard to control the animals that come at night'.

Rammam's farmers who do not resign to the 'no solution' option mention fencing as a raid prevention measure. Some mention high fencing, others, concrete long walls. One farmer said plants can be enmeshed. Also, the Forest Department does not compensate for crop loss. Farmers corroborate Roshan Rai et al.'s study about compensation being confined to large mammal raids. A *Rai* farmer and six others met a Range Officer some years ago when bear raids intensified. Why they decided to do 'something about it' hints at opportunistic agency of bears. The animal knew when the crop will be 'available'. The Ranger 'showed us a manual and it said that under forest department rules, only damage caused by elephants will get compensated and other animals are not even mentioned. But there are other wild animals that raid crops and cause a lot of damage especially in the hills. Farmers stopped reporting such problems'.

In these depredatory and depriving circumstances, what wildlife concern or contempt do people reveal? Of significance here is the forest. It played an important role in gauging wildlife concern.

Less than half the Lalung farmers we spoke to showed concern for wildlife. They conveyed animal concern with reservations or reasons. Some, like a *Dukpa* elder, felt sorry for animals. Others, like a 50-year-old *Sherpa*, justified the wildlife movement. 'They keep coming and going because we stay in the forest area'. But others conveyed animal concern with reservations or reasons. 'Forest for both', said a middle-aged *Dukpa*. 'Animals need to be protected'. 'It feels good to see them and they trouble a little also'. As against forests needing protection, they played other protective roles for farmers. A 36-year-old *Rai* farmer says: 'It will be good, if the wild animals will be there in the jungle then it will be good for us also. There will be protection of the forest also'. Wildlife are of instrumental concern for farmers here. Manifestly as forest protectors, and latently as crop non-raiders.

A *Rai* elder felt that forests are 'important for the animals as well but we have lots of them in here only'. Animals appear to not leave him and his fellow farmers any choice,'forest need be conserved for us as well as the animals but the animals create a lot of problems for us. We should think about how to save them without killing them. We cannot kill them also'. Not being able to kill boars or rabbits appears a moral thought, but it actually speaks to a legal constraint. The *Rai* elder says: 'What solution can I say? The forest and the wildlife domesticate those animals and we cannot do anything about it, we cannot kill it'.

But women do not beg the question. 'Wildlife should not be protected. Should be killed', a 52-year-old *Subba* lady observed with anger. The 'rabbits, wild pigs and others; we don't get to eat anything because of them'. Her goats and a cow were depredated. 'All we can do is to put a trap and kill those animals'. 'The deer, wild pig and the rabbits; if I could then I would have killed them', she says laughing. A 28-year-old *Tamang* woman talks about electrocuting monkeys that raid rice and flour.

A 60-year-old *Rai* woman explains this anti-animal stance. 'There are some wild animals who harm the people and destroy the fields' 'but some of them stay in their limits and feed there only. But the people don't understand that they breed, feed and stay in the jungle'. 'The people are like that only, they cannot think beyond killing and eating them but they should think. The wild animals, they are good'.

'Subjective' wildlife opinions aside, farmers afford 'objective' reasons for animal behaviour. It is a stretch to term these local theories as adaptive preference. They have some merit as efforts to reason with why animals pillage peas and maize. The *Rai* elder argues: 'If they have a proper shelter then maybe the animals will not come to our area'. 'If there was a big jungle over there then maybe they would not have come here'. Forest cover is not enough for animals and that is why they raid, is the logic. A middle-aged *Rai* woman corroborates that in a 'good jungle the birds and animals living there will not have problem of finding food'. Wildlife won't enter the village.

A younger *Rai* woman provides a more vindictive agency to animals. When they set up traps often, animals caused damage. Now they do not set traps much, and animals do not harm. It is good if 'we protect them'. 'The more we go after them they create that much damage'.

In Rammam, farmers are in consensus that forests are important for animals and them. They never question the animal's entitlement to the biodiverse Singalila habitat. But they also reiterate their entitlement and dependence. A young 26-year-old *Rai* farmer says: 'Yes the forest is very important for them'. 'There are some things for which we need the forest for example the ferns, bamboo shoots even the wildlife eat that'. Another young *Rai* says forests are important for wildlife and people should 'also use them'. A middle-aged *Chettri* concurs about people's and wildlife's mutual need of forests. As does a middle-aged *Sherpa* for whom 'villagers should also use it and it should also be there for the wildlife'. These 'coexistence' or 'co-dependence' stances are safe ones stated nonetheless with a sense of entitlement. But there also powerless opinions that give away adjustment and acquiescence. 'What can we say . . . wildlife, I guess, should be conserved', is a middle-aged *Sherpa* carpenter's disposition. Another *Sherpa* contemporary says animals need forests, 'after all it's their home'. 'We should regulate our activities', a 28-year-old *Rai* farmer felt.

There is also the question of dignity. A middle-aged *Rai* observes: 'Yes, it is important to conserve wildlife. People willingly do it. But then, it feels like wildlife has become more important than human life, over here. The value of humans is less than wildlife value'. Villagers obviously need the forest.

Animal concern is also contingent. Says the wife of a young *Rai* farmer, 'It is good if there are wild animals . . . but . . . when they cause so much damage then it gets very difficult for us'. The *Chettri* farmer who mentions mutual entitlement of animals and humans to forests replies to the question of forests and animals: 'See, those of us who work in the fields . . . let's say we sow 1 kilogram of potatoes then the wildlife come and damage that crop.

Almost half of it'. Wildlife do not offer them a choice to show concern, it seems. And they do not have the choice to chase or kill. The young *Rai* who mentioned the forest's importance for collecting ferns and bamboo says, 'we can't kill them because they fall under the wildlife department . . . we can't even do it'. An old *Tamang* resident recalls that previously 'hunters used to come and chase and hunt these animals. Now, it is not allowed'. A 30-year-old *Rai* begs the legal question even as his concern is contingent. As 'per the law we cannot kill animals how much ever they destroy our crops but the effort should be made to conserve them within the jungles and should not let them to allow to surpass the human settlements'.

Animal concern is also selective and sometimes complementary in gendered reasoning. What a 48-year-old *Mangar* lady said completed Lalung's elderly *Rai* woman's empathetic observation of animals living, breeding and staying in Senchal's forests. 'Yes they should not be there . . . except for the good ones like the pandas and the pheasants . . . they stay where they are supposed. But the wild boars, bears and leopards . . . come home and attack everything. They even attack people'. But why does wildlife enter Rammam? A 43-year-old *Thami* farmer says wildlife 'need dense forest and if they get dense forest they won't enter the village'. As the *Mangar* lady observes earlier: 'they weren't such a problem. When the common grazing areas inside the forests used to exist, that time everything was there in the forest. If one cow died, then it became food for so many animals. Now there are no herbivores inside the forests so they attack us'.

Concern for animals is a capability that is not just confined to other species but also the senses, imagination and thought capability. Farmers reveal more than a threshold level of this imaginative and reasoning capability. A 35-year-old *Sherpa* farmer claims the 'forest has been infested by thorny species which do not support neither ferns nor bamboo shoots and nothing. Most of the animals do not get food in the forest and enter the village in search of the food'. Further, 'Good forest is needed for them. When grazing was allowed they did not use to enter the village. Now they enter the village and cause problem for us'.

Conclusion

Do Lalung and Rammam farmers have other species capabilities? Do they have substantive freedoms to be concerned for wildlife and do or perform relationships with them? This capability can only be assessed and measured empirically and contextually. In tropics and temperate regions in them the context is de-humanized conservation policy and practice in 'humanized' habitats. Empirics are down to methodology and soliciting opinions and perceptions from people. But what people opine or perceive are conditioned by conservation contexts. And so any 'other species' capability assessment is automatically an adaptive preference investigating effort. Some indicative questions in this effort, as Lalung and Rammam necessitate, are: a)

Why does a certain demographic suffering depredation reveal concern for wildlife? b) If this is just a subjective adjustment to harsh circumstance, does revealing contempt for animals reveal a normal and objective human disposition? Both are subjective signals to habituation and adjustment to the 'objective' circumstances of depredation and disentitlement. Reasoning with wildlife moment, blaming human residence in the forest, feeling sorry for animals and saying animals are necessary and must be protected are adaptive to conservation discourse and practice that is at best indifferent and worst punitive, indifferent to loss of crop or sleep and punitive against any anti-wildlife action or stance.

The best Rammam's folk can do is state that forests are for both animals and people – a righteous adjustment to insecure habituation. Living and farming without title at the edges of Singalila, a famous biodiverse national park, is precarious and the choice for wildlife contempt is confined to one's head or home. But disenfranchised farming near a protected area need not always mean allying with depriving forces, and cheerfully enduring wildlife. You can be bold and beg the question: we can't kill them, can you? Or be bolder like Lalung's women and state that animals be trapped, electrocuted or killed. But Lalung is a 'plantation' village in Senchal, a wildlife sanctuary closer to urban Darjeeling, closer to the market. Crop loss is income loss, unlike in Rammam where cropping is largely for subsistence. With the Forest Department exercising nebulous control, the sense of indignation and entitlement here is certainly more than in Rammam. As a forest village, Rammam is more vulnerable to resettlement or eviction.

In other words, ostensible other species capabilities signal deprivation of other important capabilities like control over the material and political environment and an ostensibly deficient other species capability too signals deprivation of the material and political capability. Absence of land rights, and inability to participate in political choices 'that govern one's life', is a deprivation of much democratic consequence in India. The Forest Rights Act is an entitling and enfranchizing legislation that has not been implemented in India's northeast including Darjeeling. The rights and responsibilities it bestows ensure more than a threshold achievement of material and political capability. The consequences for 'other species' capability need to be watched. It has the potential to attenuate both wildlife concern and contempt. In assessing other species capabilities interesting possibilities to assess some other pertinent capabilities also remain. For instance, when Lalung's and Rammam's women explain why forest condition is linked to wildlife crop and livestock depredation, a more than threshold level of 'Senses, imagination and thought' capability is evidenced.

In a laborious, lived and changing Himalayan landscape like Singalila or Senchal, compassionate conservation or propagation of individual animal capabilities is impractically idealistic. 'Over-specified' as Alkire critiques Nussbaum's listing of human capabilities (Alkire, 2005, p. 35). For that instance in a laborious and extreme region like the Himalayas, where community is both spirit and necessary cooperation, liberal propagation of individual capabilities is also a matter of reasoning and debate.

References

Alkire S. (2005). *Valuing Freedoms*. New York: Oxford University Press.

Boström, M. and Uggla, Y. (2016). A Sociology of Environmental Representation. *Environmental Sociology*, 2(4), pp. 355–364.

Chakraborty, R., Rai, R. and Shrestha, P. (2012). *Human-Wildlife Conflict: Challenges for Conservation and Livelihood Security in Sikkim and Darjeeling*. Integrated Mountain Initiative-Sustainable Mountain Development Summit, Gangtok, Sikkim.

Darjeeling.gov.in. (2019). *Cinchona* [online]. Available at: http://darjeeling.gov.in/cinchona.html Last accessed on 2 March 2020

Dobson, A. (1996). Representative Democracy and the Environment. In: M. Lafferty and J. Meadowcroft, eds., *Environment. Problems and Prospects*. Cheltenham: Edward Elgar, pp. 124–139.

Fairhead, J. and Leach, M. (2000). *Misreading the African Landscape; Society and Ecology in a Forest-Savanna Mosaic*. Cambridge and New York: Cambridge University Press.

Guha, R. (1997). The Authoritarian Biologist and the Arrogance of Anti-Humanism: Wildlife Conservation in the Third World. *Ecologist*, 27(1), pp. 14–20.

Hitlin, S. and Piliavin, J. (2004). Values: Reviving a Dormant Concept. *Annual Review of Sociology*, 30(1), pp. 359–393.

Inskip, C. and Zimmermann, A. (2009). Human-Felid Conflict: A Review of Patterns and Priorities Worldwide. *Oryx*, 43(1), p. 18.

Kluckhohn, C. (1951). Values and Value-Orientations in the Theory of Action: An Exploration in Definition and Classification. In: T. Parsons, and E. Shils, eds., *Toward a General Theory of Action*. Cambridge: Harvard University Press, pp. 388–433.

Kothari, A. (2004). Displacement fears. Frontline 21.

Nussbaum, M. (2000). *Women and Human Development: The Capabilities Approach*. Cambridge and New York: Cambridge University Press.

Nussbaum, M. (2006). *Frontiers of Justice: Disability, Nationality, Species Membership*. Cambridge, MA: The Belknap Press Harvard University Press.

Nussbaum, M. (2011). *Creating Capabilities: The Human Development Approach*. Cambridge, MA: The Belknap Press of Harvard University Press.

Oommen, M., Cooney, R., Ramesh, M., Archer, M., Brockington, D., Buscher, B., Fletcher, R., Natusch, D., Vanak, A., Webb, G. and Shanker, K. (2019). The Fatal Flaws of Compassionate Conservation. *Conservation Biology*, pp. 1–4.

Parvaiz, A. (2017). *Darjeeling's Small Farmers Are Losing 40% of Their Crop to Raiding Wild Animals*. [online] Scroll.in. Available at: https://scroll.in/article/832715/darjeelings-small-farmers-are-losing-40-of-their-crop-to-raiding-wild-animals Last accessed on 2 March 2020

Pitkin, H.F. (1967). *The Concept of Representation*. Berkeley: University of California Press.

Rai, R., Ghose, P. and Shrestha, P. (2014). *Shifting Perspectives in Human Wildlife Conflict: Unheard Voices from the Sikkim and Darjeeling Himalaya*. Thimphu,

Bhutan: Human-Wildlife Conflict Resolution in the Mountains of SAARC -Success Stories. Mountain Ecology Division, SAARC Forestry Centre.

Redpath, S., Bhatia, S. and Young, J. (2014). Tilting at Wildlife: Reconsidering Human – Wildlife Conflict. *Oryx*, 49(2), pp. 222–225. DOI:10.1017/s0030605314000799

Rokeach, M. (1968). A Theory of Organization and Change Within Value-Attitude Systems. *Journal of Social Issues*, 24(1), pp. 13–33.

Sen, A. (1984). *Resources Values and Development*. Cambridge, MA: Harvard University Press, p. 309.

Sen, A. (1999). *Development as Freedom*. New York: Alfred A. Knopf.

Sen, A. (2000). *Freedom, Rationality, and Social Choice: The Arrow Lectures and Other Essays*. Oxford: Oxford University Press.

Treves, A., Wallace, R., Naughton-Treves, L. and Morales, A. (2006). Co-Managing Human – Wildlife Conflicts: A Review. *Human Dimensions of Wildlife*, 11(6), pp. 383–396.

Wallach, A.D., Bekoff, M., Batavia, C., Nelson, M.P. and Ramp, D. (2018). Summoning Compassion to Address the Challenges of Conservation. *Conservation Biology*, 32, pp. 1255–1265.

6 Our village would have been heaven

What the families of Gorkhey value

Siddhartha Krishnan and Sourya Das

Introduction

The 'mighty' Himalayas are vulnerable to tectonic activity and climate change. Darjeeling's neighbour, Nepal, suffered a 7.8 magnitude scale earthquake in 2015. In the last four years, the Darjeeling-Sikkim region has had 20 quakes ranging from 3.9 to 5.4 magnitude (Earthquaketrack.com, 2019). Earlier in 2009, cyclone Aila 'triggered fatal landslides in Darjeeling town' (Seidler and Bawa, 2016, p. 63). UNDP, Ministry of Home Affairs (MoHA) and Comptroller and Auditor General (CAG) reports in 2013 were scathing critiques of the condition of national disaster readiness 'eight years after passage of the Disaster Management Act, 2005'. Darjeeling was singled out along with Uttarakhand 'for the absence of minimal response plans and basic preparedness measures' (Seidler and Bawa, 2016, p. 64).

Large-scale programmes like Mahatma Gandhi National Rural Employment Guarantee Scheme (MGNREGS) and the National Disaster Management need explicate the links between development and risks. They can help mainstream disaster management and climate awareness in the Himalayas. 'There is tremendous scope for people's involvement in labour-intensive programmes promoting landscape security and sustainable natural resource use' (Seidler and Bawa, 2016, p. 67). But labour and cooperation in the beautiful but burdensome Himalayas is not a programmatic opportunity alone. They are human necessities of much character in a quacky, erosive and cold landscape. Ecosystems do not service people so much as people service themselves. It is expensive up there in Singalila and not just because boars and porcupines raid precious yield, but also because lugging essentials to these remote parts needs horses working for men or men working like horses. Community or social capital, like Sen equates it with, is not merely a relationship crucial for conversion of resources into quality of life. In the Himalayas, it is life itself that is constantly laborious and instantly cooperative.

India announced its National Action Plan on Climate Change (NAPCC) to mitigate and adapt to climate change in 2007. On June 30, 2008, it made the plan public. The NAPCC has seven nation-wide missions ranging from solar energy to water efficiency. However, it has one more region-dedicated mission – National Mission for Sustaining the Himalayan Ecosystem. In February 2014, the Indian government approved the mission. The Mission will assess the health of Himalayan ecosystems to enable policy formulation. Policies would sustain ecological resilience and ensure continued provisioning of key ecosystem services. Enhancing socio-ecological resilience needs an understanding of interactions between human and ecological functioning.

The Mission must pay attention to idiosyncratic histories of service co-productions such as of Gorkhey, a small pretty hamlet in Singalila. Investments enhancing cooperative conservation policy and practice can only follow. At regional scales, the idea of services assumes rhetorical value, that is the unilateral servicing of human needs by forests. In intimate scales, the services idea unravels in more empirical and democratic manner. Sustaining a forest citizen's freedoms must complement sustaining of forest servicing of 'needs'. Fuelwood provisioning is important. But understanding community energy reasoning and preference is ecologically significant and just.

The previous chapters compared two, or more than two, Singalila and Senchal villages and hamlets. This chapter discusses the 'affiliation' and 'practical reasoning' capabilities in Gorkhey. The idea of freedom and choice is central to human functioning and flourishing. Implicated in the choices that Gorkhey's farmers make or cannot are their relations with the forest. Some old relations have lapsed, and others persist, with varying conservation implications. Conservation laws appear to coerce choice to cultivate rather than graze. As a development basis, freedom seems lacking in this microscopic setting. Or does the decision to cultivate hint at agency? Do Singalila's forests help people, or are people helping the forest help them?

We discuss community, labour and cooperation in Gorkhey as Nussbaum's 'affiliation' capability affiliates. We assess in Gorkhey if there are more than threshold levels of functioning related to this capability. This is evident from every day and disaster time cooperation. Gorkhey is a good Himalayan case study of an interesting relationship, one between Nussbaum's affiliation capability and Sen's 'conversion' clause. The clause has it that it is no guarantee that a given level of income will translate into valuable functioning. An extreme disease, cold environment figures as a conversion constraint. But the affiliation capability can weather such 'environmental' constraints in Gorkhey. If it is robust, as we shall assess, it can help overcome such conversion constraints. But Gorkhey's affiliation or community capabilities are not the state's good doing. They signal the undoing of the conservation state's capability thwarting. And the state is no evil monolith. It can thwart, but also abet. It can curtail grazing access, just like it denies road access. It can deny electricity. But it can promote eco-tourism that people earn culturally and economically from.

Here, Nussbaum's other architectonic capability of 'practical reasoning' is in ample exhibition. Their reasoning of forest ecology, on dietary or energy choices, or why tourism is important, are revelatory; one, of what people value in a conservation context, and two, how people are valuable in conservation. The collection of scarce nutritional resource from the forest for others signals affiliation. In the inability to make energy switches, their capability to conceive and achieve what is good, convenient and sustainable is denied. But their practical reasoning is manifest in natural-historical observations of manure deficiency and bamboo regeneration. Here, their practical wisdom assumes conservation policy significance. We take the reader on a valuable tour of community life in Gorkhey. We then conclude with valuable insights on practicality, agency and ecosystem service co-production. But we now lay the conceptual ground in the following section.

Community, capability and conversion

Nussbaum's central capabilities support each other in different ways. But she attributes architectonic value to two in the list, namely, affiliation and practical reason. 'They organize and pervade the others' (Nussbaum, 2011, p. 39). Let us use Gorkhey as an example. The hamlet manages to be well-nourished despite crop depredations. But people's practical reasoning for a dispensary is ignored. This situation is incommensurate with dignity. Confining a community to growing its own maize and peas and remaining indifferent to their reasoned need for allopathy is infantilizing them. Eating the maize you grow and supplementing growth eating medicinal plants sounds 'traditional'. But, if a *Subba* reasons that allopathy is also good for her children, then we respect that reasoning. Practical reasoning is about choice. This is central to the 'notion of capability as freedom' (Nussbaum, 2011, p. 39). Before we get to affiliation, that other 'architecton' (pardon the masculinity here) that practical reasoning pervades, we need anticipate a specific doubt the reader may have.

Practical reasoning

Practical reasoning sounds intuitive and obvious and also rings so close to 'common sense'. If so, why would philosophers make the obvious an organizing principle that state policy must support? Using a common phrase involving the word – 'Get practical, people' – why would being practical be a development goal? But there is more to why philosophers value practical reasoning. Being practical has more meaningful heft than its reductive down-sizing or shaming connotation. A typical use – 'You want roads and hospitals in a Protected Area? Be practical, the Wildlife Act won't allow these destructive things'. But that is precisely what people are being: practical and reasoning. Nussbaum lists practical reason as the ability 'to form a conception of the good and to engage in critical reflection about the planning of one's life'. This 'entails protection of the liberty of conscience' (Nussbaum, 2011, p. 34). Hers must be a concise statement reflecting the history of thought on practical reasoning. But there is more to practical reasoning than Nussbaum makes accessible, at least in her *Creating Capabilities* book. For our purposes, Rasmussen's 1999 paper discussed in the introduction affords complementary insight.

In explicating the substantiveness of reason, Rasmussen may well be defending people's conscience against the callousness of experts, scientific and bureaucratic and like conservationists who trivialize the troubles of an 'illiterate' mother of two forest children. The 'functioning of one's reason or intelligence, regardless of one's level of learning or degree of ability, does not occur automatically. It is something the individual initiates and maintains. It requires effort or exertion'. Further, by way of suggesting that forest needs are non-trivial – 'Effort is needed to discover the goods and virtues of human flourishing as well as to achieve and implement them' (Rasmussen, 1999, p. 11).

Against criticisms that Nussbaum's capabilities list is intellectualized and Aristotelian and over-specified, Rasmussen offers unintended relief (Okin, 2003; Alkire, 2005). This is applicable to at least where practical reason enlisted as a central capability. For Rasmussen, it is erroneous that abstract ethical principles alone determine an individual's proper course of conduct. In our case, it is not only his Buddhist ethics that dissuades a *Sherpa* from trapping raiding boars. But wildlife laws too. So, the *Sherpa* persists by virtues of Buddhist and lawful dwelling. Also, through the practical realization that his dwelling is precarious and is vulnerable to eviction. However unjust, this is still a practical matter of survival for people.

Determinate ethics is ethical rationalism for Rasmussen. It fails to grasp that 'ethics is practical and concerned with particular and contingent facts'. Practical reason is intellectual guidance of conduct. Its virtuous use forms practical wisdom. 'Practical wisdom is not merely cleverness or means-end reasoning'. It is a person's ability to discern what is morally required in particular and contingent circumstances. There are no a priori universals to weigh goods and virtues for well-being. Using practical wisdom, people weigh (Rasmussen, 1999). In isolated and stressful Himalayan contexts, there is much weighing by the community – of water, of chances of a nocturnal crop raid, of liquid petroleum gas versus firewood. Let us examine Nussbaum's other architectonic capability of 'affiliation' and how it links to community and labour.

Affiliation

Affiliation is the capability to 'live with and toward others, to recognize and show concern for other human beings; to be able to imagine the situation of another' (Nussbaum, 2011, p. 34). Protecting this capability amounts to protecting institutions that constitute and nourish affiliation. It is also the capability of having 'the social bases of self-respect and non-humiliation'. Additionally, it is the capability of treatment as dignified being worthy and equal to others. This involves non-discrimination on the basis of race, sex, caste, ethnicity or religion. Nussbaum writes that the affiliation capability pervades other capabilities. When these capabilities achieve a dignified threshold, affiliation is part of them. The 'person is respected as a social being' (Nussbaum, 2011, p. 39).

Living a dignified, concerned and empathetic life is what makes a community. The community and its local institutions are now the alternative to state-controlled or market-permeated conservation. Community has assumed prominence in contemporary development, environment and conservation literature. Twenty years ago, Arun Agrawal provided a most incisive critique of 'community-in-conservation'. Conference and workshop reports, policy publications, NGOs and donors insist on community involvement in sustainable resource use. This valorization of community is evident in power devolution, participation and cultural autonomy literature. In discussing community's role in conservation, 'scholars display a penchant for highlighting the positive'. With the consensus on the impossibility of biodiversity protection in 'hermetically sealed parks', and the questioning of 'wilderness', community has gained cultural and political capital. Much of the biodiversity conservationists want to protect is in tropical landscapes with human populations. Creation of large Protected Areas is politically difficult. Conservation has failed without the cooperation of communities who have and defend customary rights. Coupled with grassroots development, human rights and indigenous movements, community involvement in conservation became norm (Agrawal, 1999, p. 97).

Two meanings of community are conflated. One is 'community-as-shared-understandings' where relations are organic, harmonious and self-reproducing. Two is community-as-social-organization that evokes 'territorial fixity, stable memberships, regular and face-to-face interaction'. Conceptual problems lie in attempting to establish direct links between these two. Representational problems emerge from the fact that real communities 'seldom correspond to the ideal of small, harmonious, cooperative social entities' (Agrawal, 1999, pp. 100–101).

These representations continue for political reasons and scholarship, especially political-ecological, continued to essentialize and valorize community. Amita Baviskar provides a succinct strength and weakness commentary of political ecology. Its 'great strength has been its unwavering focus on issues of social equality and justice' at stake in natural resource conflicts. While power asymmetries are 'identified as crucial in political ecology's analysis, they tend to be viewed through the binaries of civil society versus state, "virtuous peasants" versus "vicious state"'. State and communities are often assumed as 'separate, autonomous entities with self-evident interests that are clearly opposed'. Anthropology and resource conflict studies afford 'more complex representations'. But the 'continued fondness for singular narratives to explain natural resource struggles is inexplicable'. Such analysis 'takes at face value the simplified political representations that social movements must generate in order to cohere' (Baviskar, 2003, p. 5052).

Similar valorization is evident as an abstract universal in Nussbaum's affiliation capability. Many things that critics say India's stratified and hierarchical communities do not practice are affiliation indicators for Nussbaum. To recall, concern, empathy, self-respect, dignity, non-discrimination on caste or gender basis. In an unjust society where patriarchy and caste dominate, two or more capabilities of a person may conflict. Or any one or more of a person's excessive capabilities may mean another person's capabilities suffer. So, a less than threshold level of a 'lower' caste person's affiliation capability because of discrimination may have what Wolff and De-Shalit (2007) call 'corrosive disadvantage' on other capabilities (Nussbaum, 2011, p. 44). Or a more than threshold functioning of an upper caste patriarch's political and material environment capability can well corrode a 'lower' caste women's life and bodily health capabilities.

Exploring these possibilities means engaging with post-colonial critiques of capabilities. They ask what if an individual finds value in harming another? There is much value in pursuing such lines of enquiry in the more stratified societies of West Bengal's plains. Or even in Darjeeling town. But in Gorkhey, where people have no material environment capabilities and almost no political environment capabilities, corrosive potential is dim. Affiliation is a 'fertile functioning' that for Wolff and De-Shalit promotes other capabilities and functioning. This is notable in a literal way in Gorkhey's past activities and practical reasoning. We are not ruling out any tensions and animus between Gorkhey's *Chhetris*, *Tamangs* and *Sherpas*. And we must admit to not sensing any overt caste, class or ethnic conflict during fieldwork. We only note people's commonality as a Protected Area precariat. And the cooperative necessities of such precarity in a lush but laborious landscape.

Some other fascinating conceptual and political parallels remain when we compare critiques of community with critiques of Nussbaum's capabilities. In this case, the affiliation capability in particular. Sabina Alkire says Nussbaum's list is over-specified and paternalistic. Nivedita Menon says Nussbaum is an over-reliance on the role of the 'rational' and 'benign' government and its constitutionality (Menon, 2002, pp. 154–164). Frances Stewart observes that Nussbaum's approach reflects 'western liberal' values and is 'paternalist' and 'colonial' (Stewart, 2001, pp. 1191–1192).

Community too has been an over-specified concept in a conservation context. Way too much redundant and repetitive information pervades this field. NGOs, donors and scholars have also been over-optimistic of what communities are capable of in conserving biodiversity. Additionally, community fits the 'timeless universal value' bill too as it is a 'warmly persuasive word' as Agrawal quotes Raymond Williams. It is a word that has through history 'moved in and out of fashion'. Its referents are complex, heterogeneous and we cannot measure or define them with ease (Agrawal, 1999, pp. 92–94). But to Nussbaum's credit, she does list this immortal warm universal. In her affiliation capability, community is a referent, as are concern, empathy and non-discrimination.

But there are tensions in measuring Nussbaum's affiliation capability in community-conservation contexts. The affiliation capability is about what an individual can be in a community or what he or she can do as a community being. In an environmental or conservation context, this means that it is good if a *Tamang* individual has collective connections and concerns. But in a community-saturated conservation context, where the participation in community activities can develop collective interests, 'divergent interests, beliefs, and identities' will be aligned. The 'disjunction between' the *Tamang*'s 'individual and social preferences' obliterates (Agrawal, 1999, p. 101). What about social selves and sociability? the philosopher asks. What about the individual and individuality? the political scientist asks.

Our critical purposes here are not so much to 'devalorize' Gorkhey people. We did not set out to celebrate their conservation virtuosity in the first place. (We note later how home-stay opportunities are not distributed among families in Gorkhey.) Any virtuosity we engage with is their practical wisdom as implicated in practical reasoning around conservation constraints, development opportunities and ecological change. While experiences and observations of ecological change signal practical reasoning, such change also signals agency. For Sen, agency is about the 'actor' and the change she makes happen. Some 'external criteria' may be used in judging her actions. But her actions are to be judged by her own 'values and objective'. Sen is concerned 'with the agency role of the individual as a member of the public and as a participant in economic, social and political actions'. This involves direct or indirect involvement in 'individual or joint activities in political and other spheres' (Sen, 1999, p. 19).

In Gorkhey, the 'collective' agent i.e. agent in a collective, or collective agency, is pertinent in understanding ecological change and well-being consequences. Joint activities have induced indirect ecological and human well-being. As an agent then, the Gorkhey farmer is a member of the forest public or public forest. He participates in economic and social activities that influence ecological functions.

External ecological criteria judges and agrees that Gorkhey's individuals co-produce bamboo provisioning services (see Figure 6.1). Co-production of services forges stronger links between the ecosystem services and capabilities frameworks. The latter, in emphasizing freedoms of community members, considers them more as active thinking agents than passive folk whose needs are provisioned by the environment.

Co-production imparts agency by crediting people with intended or unintended roles in ecological processes. The idea of co-production of services is a critique from a labour and rights-based perspective. It assumes capabilities equivalence in terms of the 'conversion' aspect that Sen (1999, p. 74), Robeyns (2005, p. 99) and Polishchuk and Rauschmayer (2012, p. 107) emphasize. Environmental diversities like climate, including flooding or temperature, influence 'what a person gets out of a given level of income'. These environmental conditions are like social conditions for Sen. They

Figure 6.1 Bamboo, a competition

Source: Annesha Chowdhury.

influence conversion of incomes and resources into quality of life. Crime in a locality is a social conversion deterrent. Pollution too is a socially influenced deterrent to convert resources into well-being.

Individual conversion factors too, remain. A person may possess public goods, be they income, rights or resources. But personal characteristics dictate how these goods convert into valuable ends. Primary good prosperity stops mattering after a point for a disabled or aged person. They both have lesser opportunities than able-bodied or younger persons in leading normal or good lives. The aged and disabled alert us to coupling or intersectional disadvantages between 'income deprivation and adversity in converting income into functionings' (Sen, 1999, p. 88). The disabled and old have a lesser ability to both earn income and convert it into capability. This is 'real poverty' or 'capability deprivation'. This is a crucial concern in evaluation of public action to assist groups with 'conversion' difficulties (Sen, 1999, p. 88).

These diversities and variations then are conversion factors for Ingrid Robeyns – the personal, social and environmental. They influence the 'relation between a good and the functionings to achieve certain beings and doings'. These are 'living' circumstances of a person we need to know. Because 'knowing the goods a person owns or can use is not sufficient to know which functionings he/she can achieve' (Robeyns, 2005, p. 99). Besides plurality of functionings, Sen accounts for human diversity by focussing on personal and socio-environmental conversion factors in addition to the social and institutional context affecting the conversion factors.

So, we have biodiversity as potential human well-being circumstance and environmental diversity as potential human ill-being circumstance. How can we reconcile these two circumstances? We can argue that whatever be the 'hostile' bio-physicalities, that is climate, epidemiology of an environment; it can still be a biodiverse one that nurtures well-being. Singalila can get cold in winter and the *Sherpa* there needs to spend more income than his *Siliguri* counterpart to stay warm. But this disproportionate income spending can somehow be compensated by the biodiversity services that the forest affords. The forest bees pollinate his peas, right? Yet, this is no substantive reconciliation. A cold and landslide-ridden environment may be a biodiverse one. But this biodiverse environment can still be a hostile one that thwarts income generation or conversion of income or resources into functioning or flourishing. This is not because they are a hostile biophysical environment. But because they are a hostile policy environment. The forest and wildlife acts prohibit and punish. Let us push the disability analogy here. As against physical infirmity we have legal infirmity. Living in a well-serviced landscape whose biodiversity services may be robust can still be a handicapping or disabling circumstance.

Paraphrasing Robeyns, knowing the ecosystem services a person has access to is not sufficient to know which functionings she can achieve. By law, you cannot convert a Protected Area, and you cannot convert in a Protected Area. This is an institutional context affecting service conversion into functioning. But this context does not deprive all capabilities in Gorkhey. Gorkhey may be deprived of its material and political environment capabilities. But its people's affiliation and practical reasoning capabilities are palpable in what they value and achieve in terms of energy, nutrition, education and tourism.

Before the reader enters Gorkhey, there is yet one other thing related to conversion. Ecological economists argue that certain ecosystem services are themselves conversion to well-being factors. Our bigger grouse is that services are themselves tough to convert into functioning or flourishing because of legal infirmity. A *Thapa* can find it prohibitive to convert services into income or use income from co-produced service. The economists do approximate such institutional impediment. But we need to assess what the individual or collective co-production of services means in this service-as-conversion argument.

For Polishchuk and Rauschmayer (2012, p. 106), a general linking of food provisioning service and material life is inadequate in evaluating well-being in ecosystem contexts. This conceals personal and social factors influencing ecosystem services use. People have individual ways to convert cultural and provisioning services into well-being. It is crucial to consider the context and examine the process of 'conversion of ESS into well-being components'. It is crucial to specify what services are and how they differ from conversion factors. Robeyns provides scant guidance on this. Provisioning and culture influence capabilities but 'not in a one-to-one relationship'. Medicinal plants are converted into well-being through health and beliefs, for instance. Cultural services of a recreational park differ in their well-being impacts on people of different social strata (Polishchuk and Rauschmayer, 2012, p. 107).

Ecosystem services must also be explored as conversion factors. Regulating services affect and maintain climate and quality of soil, for instance, and qualify as conversion factors. They impact how people use available resources. A farm plot or 'good' 'does not in itself enable a pre-defined range of capabilities' for a farmer (Polishchuk and Rauschmayer, 2012, p. 108). But, proximate regulating ecosystem services like erosion and flood prevention add value to the land. They enhance the farmer's capabilities, that is being able to farm and harvest high yields. Here, regulating ecosystem services convert available goods into actual capabilities. They can be considered environmental conversion factors. Likewise, cultural ecosystem services can also serve as conversion factors. For instance, they 'create conditions for people to put the available resources into use'. They provide an environment for jogging, painting and poetry (Polishchuk and Rauschmayer, 2012, p. 108).

The economists treat Sen's 'environmental' conversion factor as enabling rather than constraining as an attempt to a novel ecosystem service conversion parallel. Even if regulating and cultural ecosystem services act as conversion factors, social conversion factors like policy can intervene. But legal thwarting of a *Chhetri*'s cultivation of an ecosystem 'regulated' farm is not the only 'capabilities' issue at stake here. Nor is the regulation of recreational use that an ecosystem culturally services. At stake here is people's agency. Ecosystem services are co-produced. People's labour is involved in anti-erosion measures. Gorkhey is pretty for the tourist because it is a laboured landscape.

Gorkhey: community and practicality

The Gorkhey forest hamlet is situated in the north forest range of Singalila National Park (see Figure 6.2). It is bordered by Sikkim. It falls under the revenue jurisdiction of the Darjeeling-Sadar, sub-division of Darjeeling district. The Rammam River flows along Gorkhey's east side. It separates West Bengal and Sikkim. The Gorkhey River flows along the village's west and joins the Rammam River. Samanden, which we discussed in Chapter two, lies to Gorkhey's southwest. The nearest Sikkim village is Bharang, 5 kilometres away. Bharang enjoys all-weather road connectivity. Every day, there is a service of one shared vehicle to Jorethang, a Sikkim town. It takes almost three hours to go to Jorethang from Bharang. Gorkhey has no road network. People use forest paths to go either to Bharang or to Shepi, another

Figure 6.2 Gorkhey forms a part of the famed Singalila trekking route

Source: Generated by Abhishek Samrat.

West Bengal village 24 kilometres away to access motor vehicles. People either walk or ride on horses to reach nearby villages.

There is one primary school at Gorkhey. For upper primary, secondary and higher secondary education, people send their children to different schools in West Sikkim and Darjeeling. Gorkhey does not have electricity connection. Some households have installed solar plates for generating electricity. This is for basic lighting and charging mobile phones. During monsoons, sunlight is at a premium and solar power is inadequate. Some years ago, a mini-hydroelectric unit was installed in the village with the Forest Department's assistance. But it went defunct due to the lack of maintenance. The Forest Department is reluctant to provide the 'no-objection' certificate for an electricity connection line. Wildlife and forest laws do not permit power through forests.

Buddhists and Hindus are Gorkhey's demographic. People speak in Nepali. The main occupation of people is agriculture. Every household has a small amount of agricultural land that they cultivate without title. Maize, potatoes and peas are popular crops. Farmers generally cultivate two crops in a year. The temperature during the winter falls below zero degrees Celsius and lands remain fallow. Livestock rearing, eco-tourism and labour form other livelihoods strategies. People rear cows, goats, pigs and poultry. Gorkhey is a popular spot in the famed Singalila trekking route.

There are three tourist lodges or 'home-stays' in Gorkhey village. Village folk also earn some income through the national rural employment guarantee schemes like the Mahatma Gandhi National Rural Employment Guarantee Act (MGNREGA) implemented by the forest or *panchayat* department. The Forest Department also provides occasional wage employment for plantations, foot-track and drains. People carry their agricultural and livestock products to markets in Bharang or Rimbik either by horse or by foot. They buy rice, wheat, vegetables, meat, utensils, clothes and other household requirements in these bigger villages and towns. Gorkhey houses are mainly constructed with timber, stone and asbestos. There is an informal governing institution in the village named '*Gram Samaj*'. All the adults of the village are members of the '*Gram Samaj*'. The main function of the '*Gram Samaj*' is to help during any crisis.

Vitamin and protein sourcing: a collection ethic

A household's breakfast and dinner are typically a combination of carbohydrates like maize, rice and potato; proteins, specifically *dal*; and ferns and other leafy greens. Gorkhey families grow their maize. But as a carbohydrate source, rice has substituted maize. Rice is bought from Jorethang, Sikkim. Maize, besides being a carbohydrate source, was also a cash source, but has now become very vulnerable to crop raids by boars. But its labour-intensive nature along with shifts in generational dietary preference have contributed to maize's displacement as the chief carbohydrate source. Potato is the main starch source.

A few families grow ferns and other leafy greens in their kitchen spaces or farms. The collection of such vitamin sources from forests is now an opportunistic activity. As a *Rai* woman observes, 'there is some, during the monsoons. If we do chance upon them, then we bring them home'. 'If we do go and find them on the way then we collect them'. When someone finds leafy greens in the forest they collect for others too. An old *Thapa* says, 'personally I don't get time to go to the jungle and collect those items, but yes sometimes I have only if someone gets and share it'.

Speaking of accidental or opportunistic leafy green collection from the forest floor, the community has adapted practically to the travails of scarcity and distance through expedient procurement arrangements. When people go to Jorethang for purchases, they buy *dal*, rice or chicken for themselves and many others. A middle-aged *Sherpa* tries to explain what dictates meat purchase quantum: 'we don't get it here, we have to go Sombarey, Jorethang to bring the meat'. 'Sir it's like this, if someone goes they will bring 25–30 kg for everyone'. And to a question of how many times a month is meat consumed, 'Sir it's like this if you go and buy you have to buy like 5 kg'.

At one level, this signals an 'affiliation' instance – living with and towards others. At another level this instances practicality. It is reasonable that when people visit a distant market place, they buy for their brethren which equates with a moral economy of shopping and is also reciprocity necessitated by roads that conservation laws do not permit. Put differently and unkindly, necessity breeds community. But community assumes a more reasoned and expansive form. Gorkhey imagines the situation of the ecological 'other', namely, the ecosystem. This imagination is a reasoned one. Notions of what is good for the forest and them pervade discussions of nutrition and energy choices.

Practical reasoning on nutritional and energy quandaries

A nutritional quandary also appears to have visited the Gorkhey landscape. While forests seem to have become deficient in the provisioning of leafy green and edible ferns, their capacity to provision bamboo, a raw resource of much domestic and farming utility, has also drastically declined. Singalila had a pastoral past. The cow sheds in the forest are a secure, albeit slushy, marker of past forest access and use. People reminisce about urine, dung and manure. It is almost as if Gorkhey has a past when its farmers enjoyed literal fertile functioning. Forest and fields were manured and fertile and crops, cows and humans flourished. People also had some material control capability. They had rights. A *Chhetri* elder when asked about land rights says: 'No we don't. But in the earlier time we have been keeping cowsheds in the forest. At that time, we had temporary *patta* for grazing. For this we were paying some amount to the forest department'.

Because of 25 years of restricted grazing, people reason, the forest floor is manure deficient. This impacts bamboo (*Malingo*) regeneration. A *Chhetri* says why: 'Since government has taken over the forest they have removed all the cowsheds from there and made it into a national park. But before that cows were free in the forest, they used to eat these bamboo shoots and defecate there so there was a lot of manure there so this helped them grow abundantly'. A *Bahadur* lady alerts, 'one thing if you don't have animals this forest will be destroyed, cultivation will be destroyed and even man. Because we need cow dung for manure and even cow's urine. Shoots have stopped growing'.

Sherpa families reason around the forest's provisioning capabilities. Forest resources like soil, mushroom and *ningro* 'need manure . . . if there is no manure they will not grow'. Or, 'see previously there was lots of cows and cow sheds so due to the cow dung and cow urine which acted as a natural manure there was availability of ningro, tusa and neetle leaves . . . but now there is no cows or cow dung and that's why there is a decrease in the jungle vegetables, even though there are still jungles'.

A middle-aged *Thapa* balances out the ecologist's complaint that cows browse forest species saplings. 'When it was not wildlife, villagers had cattle sheds. The cattle used to graze in different parts of the forest. The urine and cow dung used to fertilize the soil. Now, the cattle shed is no more. So, the availability is decreasing. On one hand, the cattle used to eat up the saplings thereby affecting the forest badly; on the other hand, these cattle used to fertilize the forest soil'.

Manure shortage has also meant decrease in farm soils. The solution accentuated the fertility crisis. A septuagenarian *Chhetri* recalls: 'somebody suggested us this manure sulphur . . . for one or two years crops grew well after that it destroyed the soil and to revive the soil we need another kind of manure or fertilizer which we cannot afford. . . . So, I think it's because of this harvest is low'.

For a young *Sherpa*, 'one of the reasons for less maize production is because of the lack of cow manure. If we practice shifting agriculture then potatoes, beans and maize productivity grows up but at the same time cow manure should also be mixed up at the same amount'.

After the grazing ban, keeping even a single cow is a laborious and choiceless one. A middle-aged *Sherpa* says with resignation: 'sometimes we don't want to keep it but what to do, we don't get manure for better farming. That's why I'm keeping one cow with lots of pain . . . or their grass I have to go up to the place where you can see the fog, we have to go till there . . . if we go at 6 in the morning we will come back by 8 or 9'.

Energy quandary

You need to be physically able in a legally disabling protected area like Singalila. A decent society's job is to develop internal capabilities through education and other resources to enhance physical health. Distinct tasks of a decent society are to produce internal capabilities and not block avenues that offer opportunities to function according to these internal capabilities (Nussbaum, 2011). But in Gorkhey, physical health is expended in seeking resources. Internal capabilities produced and developed without the state's support have their avenues cut off literally. There is no auto-mobility. Children, men and women walk out of sheer necessity. For trekkers it is adventure. For local folk, it is adversity. And ironically, trekkers intensify local yearning and reasoning for roads. Gorkhey families prefer roads to transport liquid petroleum gas (LPG) cylinders as energy source to cook for trekkers and themselves. Trekkers prefer forest paths to use their energies and chances of them wanting an exotic firewood-cooked meal are high.

Significant distance from the nearest town and market has meant that people continue to collect fuelwood. Transporting LPG cylinders is laborious. Families prefer LPG for many reasons, but convenience is the practical reason. A *Chhetri* wife says, 'Gas would have been better. It saves utensils from burning'. The septuagenarian *Rai* observes his wife's choice, 'For her gas is convenient'. A *Rai* elder conveys a no-brainer sense, 'if we get gas then why to use wood fire?' With 'home-stays' gaining occupational and aspirational popularity, these enterprises were one way of soliciting energy choices and reasoning. A middle-aged *Chhetri* mother says, 'if wood is not available then I think gas would be better'. However, procurement difficulties dampen her choice, 'the department of forest has been allowing us to collect fallen, broken, rotten trees and woods so it is not a problem for us and even if we use gas it comes with a lot of problems like where to get the gas from, how and where to refill it'.

Further, heating is a winter necessity – 'and it's cold here so wood fire helps us with fight the cold as well'. Restocking fuelwood is easier than refilling LPG, it seems. Head-loading twigs is less laborious than back-loading LPG cylinders. The young *Sherpa* prefers firewood for the reason 'cooking food in firewood is much easier, for gas cylinder it will eventually be a problem for us to refill the cylinder'. For a home-stay, LPG were convenient but 'a little affluent people use gas cylinder poor people like us used firewood'.

Tourism is good, tourists are good

What then is tourism value in Gorkhey? Do people conceive and critically reflect on it as good? In a tourism context, how do Gorkhey folk interact among themselves and imagine the others' entrepreneurial situation? Further, 'Green Gorkhey', 'beautiful village', 'A Lesser Known Beauty' are some typical online tourist platitudes. Such 'beauty' and other aesthetic and recreational opportunities for tourists are considered in conservation circles as a 'cultural' service of the forest ecosystem. But what are the roles of Gorkhey's families in production and delivery of this cultural service (Krishnan, 2017, p. 39)?

Residents occasionally provide guide and porterage services. People also let out horses for carrying tourist luggage. Home-stays or other families who cook meals for tourists also benefit from them. A *Chhetri* woman mentions these benefits: 'Yes, if they don't have their own vegetables they come and buy from us. Some wants eggs, some chicken so if we can do that and provide it to them it will be beneficial for us as well'. Not all families offer home-stay facilities, and while some look forward to offering such services in future, a few others are constructing extra rooms with an intent to let out. As the *Chhetri* says, 'if we get a lot of customers and it runs well then we would also want to do that . . . it will help us economically even if we stop cultivating'. But tourism can only benefit families with physical space and manpower. A mid-aged *Thapa* women observes of tourists, 'yes, it is good as they help in economy'; and of home-stays, 'no, I don't want as I stay alone, I don't have anyone to help'. For some like the *Rai* elder, 'We don't have a hotel or home stay so it really doesn't matter'.

In reasoning about the environmental and cultural implications of tourist presence, people are positive and approving. Foreign and Indian tourists are observed to be more environment-friendly than local people in their non-littering and litter-picking habits. The *Chhetri* lady says confidently, 'as far as know tourists who come here from other countries do not dirty the place in fact they pick up trash if they see them lying on the ground or forest. They search for dustbins to throw these trashes. It's our local people who eat and throw plastics anywhere and everywhere'.

Tourism is considered an alternative to agriculture that is currently stressed by wildlife raids. However, some families feel that it will be difficult to offer both home-stay services and cultivate. Tourism revenue also enables children's education. And Gorkhey's families value education. A *Sherpa* father disagrees that tourism is a social and cultural threat: 'it doesn't affect much, tourist coming is better for everyone cause because of them we will be able to educate our children'. A young *Chhetri* wife mentions village and family benefits: 'I think that it's good that they are coming. It's for the development of the village . . . yes I want to open home-stay, I will be able to feed my children'.

A brief education detour here is in order. People reason that education and employment are better options for their children than cultivation. Cultivation is not yielding crop or income because of wildlife raids and declining fertility. We do not deal with wildlife depredation in this chapter. Chapter five discusses capability deprivations because of depredation in neighbouring Samanden. The depredatory realities are similar in Gorkhey. Samanden's schooling and health travails are also discussed in chapter two. These travails are easily neighbouring Gorkhey's travails.

'I want to give them good education. . . . No, I don't want them to do cultivation', argues an elder *Chhetri* lady. A *Sherpa* says, 'It will be better if they take education. Farming and staying here are difficult. They can afford that'. Another *Chhetri* farmer is even willing to relocate, 'If the government provides the basic necessities like electricity, education, and some amount of money we are ready to go'. A retired farmer even spent the meagre retirement benefit he received on his children's education. Others like a middle-aged *Rai* mother reveals an ample capability combined of practical reasoning and affiliation: 'I want to educate my children so they can be at least a teacher. If there will be a school in my village then my son can be a teacher there after his education. My dream can be fulfilled'. But along with other quality-of-life components, Gorkhey becomes a dream itself else for a younger *Chhetri* wife, 'If the government established schools, health centre, provide electricity then our village would have been heaven'.

Living in a PA can be disabling, we had argued. But a *Rai* wife disagrees about location in the Singalila National Park being a disadvantage. 'I feel it has been an advantage compared to other villages. Even those who do not have jobs can do small jobs or work. People have started with home-stays and hotel for tourists. It's because of this that we don't face so much issues or scarcity'. This could be a matter of adaptive preference or habituation to stressful circumstances. A question of 'so what if grazing is banned, infrastructure is denied and wildlife raid our crops? The forest department allows eco-tourism and so we think we are happy'. But it could also be a matter of practical reasoning capabilities – reasoning the good and critically reflecting and planning one's life.

As opposed to the presumption, tourism is not a practically reasoned good thing. Here is a *Sherpa* elder questioning the affiliation capabilities of home-stay owners. 'I think there should be a good co-relation among the social groups and strong bonding among the individual'. People 'are not complimenting each other'. These days 'tourists are being attracted with our area and many lodges and hotels are opening government too are facilitating and suggesting rotation of home stay for tourist so there can be the equal distribution of economy'. But for instance 'down below our house only one lodge is occupying this opportunity so this has not been sparsely distributed'. But the *Sherpa* agrees that, 'yes, for me tourism should be encouraged in our village'.

Conclusion: community, concern and co-production

The suspicion that an ecosystem's services must largely be co-produced pays off in fine-scaled assessments of human-nature interactions. Considering that the notion of Singalila servicing Gorkhey's domiciles arises from what choices the latter are able or not able to exercise in relation to the former's resources, it is difficult to attribute singular agency to the ecosystem. For the fact of families not being able to procure LPGs, their preferred energy source, ensures their dependence on fuelwood. It is only because of this that Singalila's fuelwood provisioning assumes significance as a service.

In the past, Gorkhey's sizeable cattle herds grazed the forests. Intense forest grazing has, post–Wildlife Sanctuary notification in 1986, been replaced by settled agriculture. Our ecological and participatory assessments reveal that forest dependence today is largely for fuelwood and fodder collection and occasional forays for leafy vegetables. Despite moderate forest dependence today, many plant species, including bamboo, do not regenerate optimally in areas where grazing and fuelwood extraction had been intense in the past.

Gorkhey's farmers offer natural historical insight in their practical reasoning over bamboo decline. This signals their 'senses, imagination, and thought' capability, or the ability to think and reason 'scientifically'. The state does not promote such ecological reasoning or imagining capabilities. Despite the state thwarting people's grazing rights, they reason with ecological changes. They imagine the forest's plight. Affiliation is about community in the larger sense, that is humans and plant communities.

Once a grazer community, Gorkhey has adapted well, with aesthetic side-effects, to grazing bans. They switched substantially to terraced cultivation. Gorkhey Valley, a beautiful medley of farms, homesteads, streams and forests, is a laboured landscape. The aesthetics that the tourists appreciate strongly signal agency and labour. Home-stays, trek guides and horse renting services, along with meat and vegetable provisioning, make the recreational roles of Gorkhey families significant. It is in this sense that the cultural services of this part of the Singalila ecosystem are co-produced.

Co-production of services is one way to forge stronger links between the ecosystem services and capabilities frameworks. The latter, in emphasizing freedoms of community members, considers them more as active thinking agents than passive folk whose needs are provisioned by the environment. Co-production imparts agency in a way by crediting people with intended or unintended roles in ecological processes.

Singalila's forests theoretically provide regulation 'soil erosion' services to Gorkhey's fields which is an enabling conversion factor. But the farmer's labour is manifest in terraced cultivation, a soil conservation practice that needs human physical effort. If a healthy ecosystem's regulating and cultural services are conversion factors, so is a healthy human with internal capabilities. Let alone the ecosystem's regulation and provisioning services, what about the farmer community's provisioning manure services to forests? Here, any bamboo provisioning services need to be acknowledged as being co-produced. The firm natural historical belief in Gorkhey is that the grazing ban has denied crucial nutrition input to the forest floor in the form of manure. This has affected bamboo regeneration and is further evidence that timber provisioning services are in fact co-produced.

For the National Himalayan Mission that seeks to ensure the continued provisioning of ecosystem services, an appreciation of idiosyncratic histories of co-production such as Singalila and Gorkhey can help invest in conservation policy and practice that is collaborative. At regional scales, the idea of services assumes rhetorical value in terms of human needs being met majestically by forests and rivers. But in intimate scales, the services idea unravels more empirically and democratically. To enhance resilience of ecosystems, one first needs to understand, relationally, socio-ecological interactions. Sometimes, sustaining and enhancing a forest community's freedoms can complement or go further than just seeking to sustain the forest's servicing of felt needs. Fuelwood provisioning is important, but understanding community reasoning and preference about energy sources is much more ecologically significant and socially just.

Capabilities philosophy, verified in this Himalayan hamlet, yields interesting well-being and conservation insight. The collection and sharing of scarce nutritional forest resource signals community affiliation. In Gorkhey's inability to make energy switches, denied is the practical choice of what is convenient and sustainable. Forests suffer as firewood gathering continues. In observations of manure deficiency and bamboo regeneration, their practical reasoning and natural historical capabilities are revealed. Ecosystem services, tangible and non-tangible, are both self-valuable and agent relative. Fuelwood, leaf litter, medicinal plants and aesthetic landscapes are not advantages that the ecosystem offers people. These 'services' take advantage of people's labour or productive activity in becoming 'beneficial'.

When capabilities become an evaluative well-being criterion, humans stop being passive recipients of ecosystem services. Capabilities emphasize substantive freedoms of community members to function in valuable ways. They are active thinking agents rather than passive folks who need the ecosystem provisions. Co-production imparts agency, crediting people with intended or unintended roles in ecological processes. The idea of co-production of services is also a critique from a labour and rights perspective. Some may labour more, others may benefit more.

References

Agrawal, A. (1999). Community-in- Conservation: Tracing the Outlines of an Enchanting Concept. In: R. Jeffery and N. Sundar, eds., *A New Moral Economy for India's Forests? Discourses of Community and Participation*. New Delhi: Sage Publications, pp. 92–108.

Alkire, S. (2005). *Valuing Freedoms*. New York: Oxford University Press.

Baviskar, A. (2003). For a Cultural Politics of Natural Resources. *Economic and Political Weekly*, 38(48), pp. 5051–5055.

Earthquaketrack.com. (2019). *Earthquakes in Dārjiling, West Bengal, India – Most Recent*. Available at: https://earthquaketrack.com/in-28-darjiling/recent Last accessed on 3 March 2020

Krishnan, S., Devy, S. and Krishnaswamy, J. (2017). *Beyond Trekker Platitudes: How Forests and Farmers Fare in an Eastern Himalayan Forest Edge*. Bengaluru: ATREE, pp. 36–43.

Menon, N. (2002). Universalism without Foundations? *Economy and Society*, pp. 152–169.

Nussbaum, M.C. (2011). *Creating Capabilities*. Cambridge, MA: Belknap Press of Harvard University Press.

Okin, S. (2003). Poverty, Well-Being, and Gender: What Counts, Who's Heard? *Philosophy and Public Affairs*, 31(3), pp. 80–316.

Polishchuk, Y. and Rauschmayer, F. (2012). Beyond "Benefits"? Looking at Ecosystem Services Through the Capability Approach. *Ecological Economics*, 81, pp. 103–111.

Rasmussen, D. (1999). Human Flourishing and the Appeal to Human Nature. *Social Philosophy and Policy*, 16(1), pp. 1–43.

Robeyns, I. (2005). The Capability Approach: A Theoretical Survey. *Journal of Human Development*, 6(1), pp. 93–117.

Seidler, R. and Bawa, K. (2016). Dimensions of Sustainable Development. *Religion, Culture and Sustainable Development*, 3.

Sen, A. (1999). *Development as Freedom*. New York: Alfred A. Knopf.

Stewart, F. (2001). Book Review: Women and Human Development: The Capabilities Approach, by Martha Nussbaum. *Journal of International Development*, 13(8), pp. 1191–1192.

Wolff, J. and De-Shalit, A. (2007). *Disadvantage*. Oxford: Oxford University Press.

7 Conclusion

Inclusion

Siddhartha Krishnan

The question here is not if conservation science and policy be inclusive of community well-being. That is an inadequate question to which the answer has always been a resounding yes. It is inadequate because of the 'how' aspect of this inclusion. We tend to reduce 'how' to entitlement. Giving rights will ensure that the community and members will be well. The forest will be well. This is a consequentialist ethic. Entitlement in conservation also enjoys a de-ontology ethic. Giving individual and community rights is right. Period. 'If' people and forests will be well is not a consequence that we include in this democratic calculus. Of the consequentialist stance, we must ask what we mean by 'well'? Of the de-ontological stance, we must ask, why is entitling right? These are inclusion matters that, to wit, we need exercise our practical reasoning capabilities! We need to include more to what being well means. We need to include more into the rights-entitling calculus. Now, to answer the inadequacy with which we began, the important question is why conservation should be inclusive of how it understands well-being.

The conservation community, including conservation biologists, ecological Marxists and political ecologists should cease being satisfied with rights entitlement. By this, I do not mean that forest rights are bad. They are good. But why are they good? Because it appeals to your liberal progressive politics? Or is it because it is a fool-proof mechanism to conserve biodiversity? The former is a well-being ally and proof itself, and in the spirit of progression we need to renew the terms of alliance and proof. The latter points to our benign beliefs. They portend unintended well-being and conservation consequences. But progressive entails being of a reformatory bent. We reform others. We reform ourselves in the course of interaction with others. We must reform our assumptions and approaches, in true democratic spirit. So, being progressive about community well-being does not end with, well, well-being as an end.

As we discussed in the second chapter, rights, ecosystem services, resources or incomes are public goods. These do not translate into valuable functioning in an automatic way. Forests rights under the Recognition of Forest Rights Act do not ensure a pan-Indian uniform well-being achievement. For that instance, even within a PA, the legislation will differ in its well-being effect. This is not to score a culturally relative point and say that the FRA is too universalizing in its intent. India is way too diverse for law to have uniform impacts. Rights, legislations can only be general with the hope that their coverage is universal. India's forests are as culturally diverse as they are biologically and diversities, like inclusion, are not just progressive or radical rhetoric here. They are keywords that in our scheme organize, for want of any other word, interdisciplinarity. These words combine and permute in more than one way.

The capabilities approach must be included in conservation scholarship. Conservation must be included in capabilities scholarship. Well-being can include more than rights, income and pleasure. More than revealed preferences or pleasures, rights and income can include freedoms as effects. Conservation and capabilities must be inclusive of diversities in human contexts and concepts. There are personal, social and environmental diversities that influence how income or rights convert into well-being. There are diversities in ways the environment thwarts resource to life quality conversion.

Diverse possibilities emerge when capabilities and conservation include each other as political-ecological and philosophical contexts. The possibilities are conceptual, methodological and political. Before we get to these suitably instanced possibilities, we must provide a quick review of major arguments that emerged from our village engagements. How useful or limited is the capabilities idea in conservation landscapes?

From the comparative case of Samanden and Lalung, and also implicit in other village case studies that follow, it appears that for a universalizing theory, capabilities fares better than utilitarianism. The latter's 'satisfaction' type measurement of forest dweller well-being is ignorant of powerless habituation to capability deprivations wrought by conservation. When a school and dispensary is absent in a conservation landscape, surveying for satisfaction over ecosystem services can only beget responses and results that lend themselves to statistical and not social significance. So from a non-specified or non-listed sense, Samanden and Lalung's people value education and health facilities. But does achieving threshold functioning of capabilities such as those Nussbaum specifies cut it in a constitutional polity? Our comparison in the third chapter of how Srikhola and Rampuria fare suggests that thresholding is all about hanging on. Residents here lack the opportunities of improving their natural environment, in other words conserve by participating. While the material environment control and affiliation capabilities are at threshold levels, the political environment control capability is deficient like most other villages we evaluated for flourishing.

This capability deprivation is also the deprivation of their Fundamental Rights. Article 21 guarantees right of life and liberty, and from Nussbaum's perspective, what the government appears to provide are thresholds or bare minimums. Further, Srikhola and Rampuria's women experience capability deprivations differently, and given how stratified Indian forest communities are, justice must be theorized intersectionally. Another point pertinent to all other villages that the authors of the third chapter make is that capabilities cannot just be expanded as an individual enterprise. It is important to shift focus of capabilities' analyses from the individual to a collective. In community conservation contexts, it is a goal of justice to constantly navigate between the individual and community.

The fourth chapter builds on Holland's critique of Nusbaum's insufficient acknowledgement of the environment, or Sustainable Ecological Capacities (SEC), significance in supporting or enabling all other capabilities in the latter's list. From a more tropical and third world standpoint, the authors, comparing all villages across Singalila and Senchal, complicate nature. It is environmentally just they say to shift from treating people as mere beneficiaries of nature to considering them as part of it. People co-produce and co-evolve with nature. The chapter is disinclined to attribute agency to nature in influencing the conversion of natural resources into valuable achievements, akin to social, political and economic institutions. Human consciousness and power disallows attributing purpose to nature. Invoking ecosystem services, whether provisioned or regulated, rather than ecosystem functioning or process, the chapter and its authors assess how people, through their labour, obtain from forests. The chapter operationalizes SEC for instance as a 'well wooded' space that people can value for timber, sequestration or recreation. But what people value may not matter, with policy-backed value determining how or what an ecosystem is managed for. For a household or individual's SEC capability, the authors recommend availability of and access to ecosystem goods as meaningful indicators to evaluate social justice. By implementing the Forests Rights Act, the state can ensure socio-ecological and political conditions for threshold-level functioning that enables the achievement of capabilities like 'life', 'health', 'control over material and political environment', 'senses emotion and thought', 'affiliation' and concern for 'other species'.

Speaking of the 'other species' capability, what is this rather bio-centric capability's worth as a well-being measure in the Himalayas? Does this ostensibly first world value have limitations in a third world temperate region? There is the limitation of probing the obvious as forest communities interact with wildlife directly and indirectly everyday and where interactions are both affable and antagonistic. But there is also the limitation of method in assessing care and love for animals – quantifiable surveys cannot with all their alleged rigour really understand if love or hate for leopards and boars is real. A forest individual's animal preferences are *adaptive*. In the fifth chapter, comparing human-wildlife interactions in Singalila's Lalung and Senchal's Rammam, we stress context and questions to ask in context. Conservation landscapes are spaces where capabilities deprivations are legislated. Here, pertinent questions are why do suffering forest folk express concern for wildlife? Is contempt that suffering forest folk reveal for animals normal and objective sentiment? Evaluating other species capability thresholds must necessarily be done in relation to evaluating material and political control capability thresholds. The entitling and enfranchising Forest Rights Act if implemented will have 'fertilizing' effects for the other species capability. Rights and responsibilities bestowed under the Act means more than a threshold-level functioning of material and political capability. The Act, along with *Panchayat* elections, has potential to attenuate wildlife contempt and enhance concern.

In the fourth chapter, we mentioned how ecosystem services, their availability and access, can serve as a concrete indicator of SEC capability. Also, we mentioned that forest people co-produce services. The sixth chapter, where households in a single hamlet in Gorkhey were compared for capabilities, demonstrates co-production instances and makes a case for making capabilities a potential framework to gauge well-being in ecosystem services assessments. Co-production for instance is fathomable in a laboured landscape that then becomes an aesthetic or recreational service for eco-tourists. Laboured efforts like building and maintaining home-stays, being trek guides, rearing horses for rent and meat and vegetable provisioning render significant the recreational roles of Gorkhey families. Service co-production can forge potential links between the ecosystem services and capabilities frameworks. Capabilities by stressing freedoms of community members treats them as active agents and not passive folk being provisioned or regulated by the forest ecosystem. Additionally, co-production allows for agency by crediting people with latent or manifest roles in ecological processes.

At the risk of being faintly repetitive let us examine, using concrete Eastern Himalayan examples, the different conceptual, methodological and political possibilities that arise if capabilities and conservation combine as political-ecological and philosophical contexts.

Capabilities and conservation: an inter-disciplinary agenda

Conceptual

Conservation activism and scholarship is a field with a heavy utilitarian influence. Despite acknowledging, for instance, complex power relations, a simplistic approach to well-being prevails. Subjective feelings and satisfaction, individually generated and extrapolated to community, are information that rights and income are working to satisfaction. Conservation biologists and quantitatively prone social scientists suffer an information limitation. Resource, income and satisfaction are their well-being informants, even as they want people to be their informants against poaching and felling. 'Satisfaction' with livelihood concessions in both entitled and disentitled contexts are well-being measures and indicators. In a disentitled context, one asks – Are you very happy/happy/little happy with your NTFP harvest quota, or with the coffee seeds the government gave you because boars will not eat them like they eat your millets? In an entitled context one asks – Now that you have rights, will you carry out your conservation responsibilities? As a critique of utilitarianism, capabilities can be productive in broadening the informational base from resources and income to real freedoms and capabilities.

PAs are a potential context to assess capabilities of the unspecified or non-listed kind (Sen) and specified and listed kind of (Nussbaum) capabilities. In many development contexts, impoverished or inequitable, income deprivations are capability deprivations. Here, social and environmental 'conversion' factors deter income or resource translation into valuable functioning. But PAs are biodiverse and resource-rich contexts where income generation or income spending avenues are curtailed. PAs offer conceptual scope to expand the 'physical handicap', an interpersonal differential in capabilities that Sen often invokes in making the point about converting resources into valuable functioning. PAs are spaces where a policy infirmary is analogous to physical infirmary in earning income or spending it in individually valuable ways.

Further, conservation contexts, especially PAs, serve as a corrective to Sen's conviction that hostile environments are conversion factors or deterrents. Malarial topical forests or cold Himalayan forests, then, influence what one can do and be with an income range. But these malarial and cold forests may not be conversion deterrents only due to their 'hostile' bio-physicalities. Exclusionary conservation laws make them hostile policy environments. Families in Singalila or Senchal hardly mentioned winter as a deprivation. Crop depredation and infrastructural disenfranchisement were the main deprivations.

Ecosystem services scholars claim regulation and cultural services are conversion factors from a capabilities perspective. For instance, Senchal provides forest paths and spots for trekking and picnicking. But capabilities scholars, especially Sen, invest in agency. One particular way in which agency manifests is through cooperative labour acts involved in co-production of services. Paths are created by people's material routines. They cleared and made spots to rest from their productive activities. Ecosystem service scholarship, specifically those who investigate ecological and human functioning in parallel, can engage with ideas of conversion and agency in the capabilities approach.

Political ecology's interests on justice and rights are compatible with procedural or listed capabilities approaches such as those of Nussbaum. Individual and conservation rights, for instance, serve as thresholds for her 'material and political environments' capability. A *Sherpa* entitled by the FRA will have a minimal 'material environment' capability. Political ecologists and 'listed-capabilities' scholars can debate on the 'architectonal' value of the material and political environment capability. If the *Sherpa* has both land and conservation rights i.e. material and political capabilities, it will fertilize other capabilities too. Including his reasoning, health and affiliation capabilities. Nussbaum privileges 'Practical Reasoning' and

'Affiliation' as architectons. However, material and political environment control capabilities are extremely crucial in conservation contexts where people's choices and opportunities are thwarted or deprived by laws. The architectonal potential of the material and political environment control capability, and Nussbaum's privileging of affiliation or community capability, lend potential scope for conservation conversations between capabilitarians and political ecologists over well-being and power relations, not to mention the significance of dignity and social justice for both these philosophically and politically oriented scholarships.

Conservationists and political ecologists say communities must have development rights. Nussbaum argues that having community affiliations is a development capability. For Nussbaum, being able to show concern for and live with others is a matter of dignity and justice. For conservation activists and scholars, communities must have rights as a matter of dignity and justice. Both valourize community and a potential area of collaboration emerges. But this doesn't drive the tension away. Critiques of romanticizing communities ask what about the 'individual' in the community? Can the *Tamang* not value city life while continuing to be a member of a forest community? Nussbaum asks what about the community in the individual? Should the *Chhetri* individual evince empathy for and participate in community? Nussbaum's 'affiliation' capability will be of interest in conservation circles that consider community a 'warm and persuasive' word. Capabilities scholars will be interested in critiques of 'individuals' being engulfed and lost in this warmth.

Methodological

Related to the conceptual scope of PAs being capability evaluation spaces is the empirical scope that forests offer for contextualizing and operationalizing the capabilities philosophy. From a non-listed or non-procedural Sen-like perspective, PAs offer empirical scope in the collection of 'valuable' information on what capabilities forest people decide are important for them and their families. From a more listed and procedural Nussbaumian perspective, forests also offer empirical scope to collect information on capabilities like affiliation and material and political environment control. But there remains a methodological rider. Perception and satisfaction surveys are typically used in well-being assessments around ecosystem service or PA contexts. These methods and their designers need be informed of adaptive preferences. Adaptive preferences pertain to subjective information that hide habituation and adaptation to stressful circumstances. The conservation scholar must be alert to subjective data passing off as objective. Here, even a structured and ordered survey instrument is no guarantee to generating objective data. Objectivity is not so much achieved here by the instrument as it is by the scholar's critical abilities and contextual sensitivity. Liking or hating tigers may be pre-coded answer ranges that conservation biologists quantify and seek statistical significance. But statistical significance is no guarantee for social significance. *P* value may not be the best indicator of what things people value and are unable to pursue.

Policy

FRA scholarship needs to consider going beyond studying entitling under the Act or the implementation of its provisions. Using the capabilities approach they can usefully ask what a right recipient is able to do and be with rights, both individual and community. Forest citizens are not passive needy recipients. They need the dignity of consideration as active reasoning agents. FRA scholars and scholars persuaded by community do display genuine concerns over the dignity of *Adivasis* and other forest dwellers. This concern over dignity can be a more expansive one that entails not just rights and income needs, but also freedoms of using choices and opportunities created by rights and income to pursue valuable things.

Sustainable development is a 'needs' based philosophy. The current generation's needs must be met in a way that does not compromise future citizens' needs. People, including forest dwellers, are considered as needy beings. Sen requests a re-focus to sustainable freedoms. It is important to take into consideration freedoms of the current generations being met in a way that does not compromise the future generation's freedoms. Forests are a good context to evaluate prospects for such sustainable freedoms. Treating and keeping people as needy infantilizes them. Enhancing people's substantive freedoms to pursue valuable things can have democratic consequences. This is true especially in the context of recent proposed Amendment to the Indian Forest Act, 1927. It bestows armed powers to forest personnel and powers to override the FRA. The Amendment mentions 'sustainable management' more than a dozen times. This understanding of 'sustainable' is a hangover of the needs-based core of sustainable development.

Glossary

Adaptive preference A state where people become so much normalized to circumstances of material deprivation and social injustice that they appear or claim to be happy and satisfied. This is generally difficult to understand in surveys and in-depth interviews when people seem to be used to their deprivations.

Alpine ecosystem A natural region or biome that does not have trees because of the high elevation. The high elevation is marked by extreme climate, which is too cold and windy to support tree growth.

Animal welfarism Also called animal welfare, an ethical position that morally allows humans to use non-human animals, while reducing or minimizing adverse effects on animals.

Bio-centric conservation A form of conservation that attributes inherent value to animal and plant species.

Bio-fencing Also live fencing of the botanical kind. Involves trees or shrubs grown along boundaries to ward off wildlife or livestock.

Capability A person's capability is the substantial freedoms and opportunities to choose between different valuable functioning combinations or the different kinds of life that she has reason to value. Basic capabilities comprise freedoms to do the bare minimum things that are considered necessary for survival and also to avoid deprivations such as poverty. Nussbaum refers to basic capabilities as 'the innate equipment of individuals that is necessary for developing the more advanced capabilities'. Thus, for instance, the capability of speech and language in a newborn is a basic capability which can then be developed.

Communitarian Communitarianism emphasizes linkages between individuals and their community. Individual identities and dispositions are considered to be shaped by community interactions. The community can be a family or a small group of people. However, communitarianism is generally understood in the broader philosophical sense, as a set of interactions among a community of people in a given geographical location, or among a community who have a shared history. It usually opposes extreme individualism and vouches for the stability of the entire community.

Consequentialist ethics The moral evaluation of human choices, both action and intent, on the basis of their consequences. Good or bad consequences must be the basis on which choices are judged.

Conservation Conventionally, the protection of forests notified as national parks and wildlife sanctuaries through strict and punitive forest and wildlife laws. More democratically, the participatory and sustainable management of Protected Areas and other forests involving the state and local communities.

Co-production Role of human labour in producing what is ostensibly an ecosystem service. For instance, in a forest's regulating service of soil erosion, humans play roles in terracing.

Decentralized development A development model that was participatory in design, implementation and monitoring of development programmes and schemes. Decentralization is the process in which the activities of an organization, especially in terms of its planning and decision-making processes, are delegated from one central authority to multiple local authorities. This removes alienation of people in decision-making processes about their development.

De-ontological ethics The moral evaluation of human choices, both action and intent, strictly on the basis of duty and what is right. The consequences of choice, even if they are morally good, do not matter.

Distributive justice Distributive justice is about the socially just allocation of goods and resources. It is usually contrasted with procedural justice that concerns fair and transparent decision making, for instance with regard to goods or resources allocation. Procedural emphasis is on process, and distributive emphasis is on outcome.

Eco-Development Committee (EDC) Similar to Joint Forest Management Committees, the Eco-Development Committees (EDC) are designed for villages in Protected Areas and their buffer zones. EDCs have two-fold objectives of wildlife and biodiversity protection, and assumption of eco-friendly development activities in villages.

Ecological Marxism An influential Marxist wing of third world environmentalism. Ecological Marxists link development and ecological crisis to political and social inequality. It also explains the ecological destruction caused by capitalism. It is built on what Marxists describe as contradictions of capitalism. In ecological destruction, the contradiction lies in the trend for capital to destroy nature, which is part of the process of expansion of capitalism.

Environmental history An historical sub-field in which nature is more than mere physical context to human events. It changes in response to human action, and also by itself. The sub-field studies the relationship between human societies and nature.

Environmental sociology A sociological sub-discipline that seeks to establish links between the biophysical and socio-cultural. Although the focus of the field is the relationship between society and environment, environmental sociologists emphasize the study of social factors that cause environmental problems, the societal impacts of these problems and also efforts made to solve them. Additionally, a lot of emphasis is laid on the social processes by which certain environmental conditions become socially defined as problems.

Epistemology Knowledge is 'episteme', and 'ology' is science or discipline. Philosophically, a field concerned with how we know what we think we know about the world. It analyzes the means of production of knowledge and also the scepticism of claims of different kinds of knowledge. It deals with the creation and dissemination of knowledge in specific areas of enquiry.

Essentialism Determinate accounts of humans and societies that espouses that they have central defining traits. Conversely, anti-essentialism espouses universal traits to humans. Plato is considered to be one of the first essentialists and he believed in the concept of ideal forms, which is an abstract entity of 'which individual objects are mere fascimiles'.

Gregarious species Animals and birds that move in groups, flocks and bands.

Historical materialism A Marxist theory of evolution of social forms. Historical pertains to the order of social form or societal unfolding. Materialism pertains to a critique of Hegelian privileging of ideas over the economy. Marx afforded primacy to the economy over the cultural development of ideas.

Historical sociology A sociological sub-field that studies history sociologically. It engages with events, the historian's staple. Historical events, wars for instance, can structure human behaviour. And human agency or will can structure events. Consequently, history must engage with sociological theory to understand structuration.

Human agency Human actions that cause change. These actions can be judged by the actor's own values and intent, or by some objective external criteria. In development, the agency role of the individual involves membership in community and public, and participation in socio-economic and political actions.

Human consciousness Consciousness is normally a given as people are conscious of themselves. Psychologically, it is the state of being aware and sentient. Humans feel and experience.

Indigene An indigenous person with native roots. Known as *Adivasis* in India, they comprise 8.6 percent of India's population.

Joint Forest Management The management of forests where the Forest Department involves local communities. Management is institutionalized and rules, roles and benefit sharing are encoded formally.

Meta-capability Since environmental conditions like ecological and climatic processes contribute to human capabilities just like shelter and nourishment do, these conditions are an independent meta-capability. They make all other capabilities possible.

Montane forest In India these forests lie above 1,500 m on the slopes of the Western Ghats and Eastern Himalayas. They have an annual rainfall more than 150 cm and an annual average temperature of 18°C. The regions achieve freezing point in the winter months between December and February.

Neo-liberal A 20th-century ideology emphasizing the value of free market competition. It comprises economic liberalization, privatization, free trade and reduction in government spending in order to increase the role of the private sector in the economy and society.

Nutrient cycling The movement and exchange of organic and inorganic matter back into the production of matter. Energy flows unidirectionally, but mineral nutrients have a cyclic movement.

Pastoral Land for grazing livestock.

Paternalism A fatherly attitude or management policy adopted by a government, company or community. As with fathers, it involves, in the guise of promoting one's own good, the usurping of individual responsibility and liberty. The issue of paternalism arises when there are restrictions by law like in the case of anti-drug legislations and wearing seat belts and also withholding relevant information of a patient's condition by doctors. It raises questions as to how people should be treated when they are not completely rational.

Perception survey Typically, structured survey instruments that solicit and rank perceptions of respondents. Perceptions themselves are unstable data not too different from a stimulus and response type situation. Perception surveys lack depth and reflection.

Political ecology A political economic sub-genre that arose as a critique of apolitical ecological studies. It politicizes environmental conflicts through an incessant engagement with power relations and justice.

Pollination cycling The transfer of pollen grains (male sex cells in plants) to the female sex cells for fertilization. Pollination can occur between plants when pollen is carried by the wind or by insects like the honeybee. Some plants self-fertilize.

Post-colonialism The historical period that represents the aftermath of Western colonialism. Also, it is a body of thought concerned with theorizing the political, economic, historical and social impact of European colonial rule around the world.

Primitive accumulation A Marxist term denoting the process through which feudal and chattel slave modes of production transformed into the capitalist mode of production. A violently displacing process where the self-sufficiency of a large number of people was destroyed.

Procedural justice Procedural justice is the idea of fairness of processes in the justice-seeking process that aims to resolve disputes and justly allocate resources.

Relativism Social groups have varied tastes, values and beliefs, and even varied ways of discerning the truth. Human life is then relative and what constitutes well-being for instance, is culture-dependent.

Reserve forest Area reserved as forest after a settlement of rights, under the Indian Forest Act, 1927.

Resource ecology The calculation and control activities that focus on the building's natural resource use, such as water, energy, landscape, waste management, soil type and groundwater.

Revealed preference A behaviourist theory of consumer choice. It emerged in the need to rescue economics from its psychological roots. Feelings and other mental states were considered unempirical and non-observable.

Romantic naturalism A very emotional naturalism, where naturalism is the idea that the individual should conform to nature.

Savannisation The domination or invasion of forests by grass species.

Self-determination An individual controlling her own life choices and leading it accordingly. Also a nation controlling its own sovereign path towards governance.

Self-realization (Aristotle) For Aristotle an excellent life has as its basis humans actualizing or realizing their potentialities. With connections to humanistic psychology, the idea of self-realization focuses on human potential and the quest for what it means to be a virtuous human.

Social capital Social capital is made up of the resources, trust and social networks, access to which empowers people to acquire advantages like well-paid and pleasant jobs. Social capital is spoken about in the sense of social relations that have productive benefits.

Social contract For Enlightenment thinkers, an ideal basis of a state. It involved, variedly, individuals sacrificing power to a sovereign in return for life and property protection; marginal ceding of such powers to the government, which can easily be deprived of such powers; and a general will that democratically participates in government. In all, the common idea is that the estate's legitimacy depends on citizens' will and wishes.

Over the centuries, philosophers such as Socrates have attempted to describe the ideal social contract and also to establish how present social contracts have evolved. Social contracts can be explicit, that is in the form of laws, or implicit, such as social norms like touching an elder's feet as a sign of respect. The Constitution is cited as an explicit example of a social contract as it gives clear indications of what the government can and cannot do.

Social structure Enduring patterns of economic, social, cultural and political relationships and interactions. These patterns influence human agency or action. Disagreement reigns over which part of the component is essential. Social institutions are usually considered good examples of social structures. The major social institutions usually used in the sociological discourses comprise family, religion, media, law, politics, economy and education. These are understood as distinct institutions that are interdependent and can collectively help society.

Socio-ecological resilience The capacity to adapt or transform in circumstances of change in social-ecological systems, particularly unexpected change, in ways that continue to support human well-being.

Sustainable development According to the Bruntland report (1987), 'Sustainable development is development that meets the needs of the present, without compromising the ability of future generations to meet their own needs'. The term 'sustainable development' is often viewed as contradictory by many economists owing to development being synonymized with economic development which cannot bring about sustainability. There is thus a movement to look at sustainable development more holistically and not make income generation and GDP growth as the only indicator of development.

Under-development Simply, any society that has not experienced economic and social modernization and associated changes. The term usually also serves as an explanation of continuing poverty of a country. Here under-development is an economic malady afflicting nations that both lack development and are exploited by other richer and more powerful nations. In development discourses, the concept of under-development is often believed to have two origins, one being the colonial period where the colonial powers exploited resources from colonies and the second is the post–World War II period or the Cold War era.

Utilitarianism In classical forms developed by Jeremy Bentham, utility is pleasure or satisfaction. An aggregating philosophy, utilitarianism is insensitive to utility distribution since its focus is on sum-totalling of utilities. In its modern forms, utility has come to mean the fulfilment of desire. Utilitarianism is a famous instance of consequentialist ethics. Here all individual and institutional choices are judged by their consequences. Utilitarians believe that the main purpose of morality is to make life better by simply increasing the number of good things such as pleasure and happiness and at the same time decreasing the number of bad things such as pain and unhappiness.

Welfarism Welfarism is an economic view that insists that goodness only be evaluated in terms of subjective utility. It concerns itself only with psychological states, and not with people's reflective evaluations. For Amartya Sen (1979), the 'judgment of the relative goodness of alternative states of affairs must be based exclusively on, and taken as an increasing function of, the respective collections of individual utilities in these states'.

References

Amartya Sen, 1979. 'Utilitarianism and Welfarism'. *The Journal of Philosophy*, 76(9), pp. 463–489.

World Commission on Environment and Development. (1987). *Our common future*. Oxford, Oxford University Press.

Index

Note: Page numbers in *italics* indicate a figure and page numbers in **bold** indicate a table on the corresponding page.

For Product Safety Concerns and Information please contact our EU representative GPSR@taylorandfrancis.com
Taylor & Francis Verlag GmbH, Kaufingerstraße 24, 80331 München, Germany

www.ingramcontent.com/pod-product-compliance
Ingram Content Group UK Ltd.
Pitfield, Milton Keynes, MK11 3LW, UK
UKHW020001200926
475702UK00015B/2412
* 9 7 8 0 3 6 7 5 3 8 8 4 2 *